Graphics Drawing Workbook

Gary R. Bertoline
Purdue University

Boston   Burr Ridge, IL   Dubuque, IA   Madison, WI   New York   San Francisco   St. Louis
Bangkok   Bogotá   Caracas   Lisbon   London   Madrid
Mexico City   Milan   New Delhi   Seoul   Singapore   Sydney   Taipei   Toronto

McGraw-Hill Higher Education

A Division of The McGraw·Hill Companies

GRAPHICS DRAWING WORKBOOK

Copyright ©2000 by The McGraw-Hill Companies, Inc. All rights reserved.
Printed in the United States of America.
The contents of, or parts thereof, may be reproduced for use with
FUNDAMENTALS OF GRAPHICS COMMUNICATION and
TECHNICAL GRAPHICS COMMUNICATION
Gary R. Bertoline
provided such reproductions bear copyright notice and may not be reproduced in
any form for any other purpose without permission of the publisher.

2 3 4 5 6 7 8 9 0 MAL/MAL 9 0 4 3 2 1

ISBN 0-07-233608-0

http://www.mhhe.com

Graphics Drawing Workbook
Preface

The *Graphics Drawing Workbook* is meant to be used with either *Technical Graphics Communications, 2nd Edition* or *Fundamentals of Graphics Communications, 2nd Edition*. However, the workbook can be used with any good reference text including *Graphics Communication for Engineers* by this author.

There are workbook problems for every major topic normally taught in an engineering or technical drawing course. Most of the problems can be drawn with instruments or sketched. A special emphasis had been put on freehand sketching in this workbook in response to the increased use of CAD in many technical drawing courses. It is expected that the instructor will supplement these problems with others from the text to fully reinforce technical drawing topics.

There are many unique features in this workbook to reflect the use of 3D and CAD in technical drawing. There are many workbook problems that focus on 3D modeling, such as the sketching problems relating to Boolean operations. There are also many visualization-related problems to reinforce the importance of visualization in technical drawing and 3D CAD.

The author wishes to thank Betsy Jones from McGraw-Hill for her efforts and assistance in getting this workbook published. The author would also like to thank my illustrators that assisted in this workbook and the solutions guide; Professor James L. Mohler, Elliot Poladian, Amy Fleck, and Nick Wainscott. Comments and suggestions are always welcome by contacting me at Purdue University, by email at grbertol@tech.purdue.edu or through the publisher.

Gary R. Bertoline
Purdue University
Department of Computer Graphics
1419 Knoy Hall, Room 363
West Lafayette, IN 47907

Contents

Graphics Drawing Workbook
Instructions
Reference *Technical Graphics Communications, 2nd Edition*

Refer to the following instructions for a detailed explanation of the requirements for all the problems in this problem book. References are to section numbers in the textbook *Technical Graphics Communications*, 2nd Edition. Use these references to assist you in solving all problems.

Technical Drawing Tools, *Reference Chapter 3*

3.1 Line Drawing 1. A. Draw or sketch 6 equally spaced horizontal lines. B. Draw or sketch 6 equally spaced vertical lines. C. Draw or sketch 8 equally spaced 45-degree lines. D. Draw or sketch 8 equally spaced 30-degree lines. E. Draw or sketch 8 equally spaced 15-degree lines. F. Draw or sketch 8 equally spaced 75-degree lines. Reference 3.5.

3.2 Line Drawing 2. Fill the space with the pattern started in each of the 6 spaces. Reference 3.5.

3.3 Scales 1. Measure each line using the indicated scale and print the numerical value using the guidelines provided. Reference 3.6.

3.4 Scales 2. Measure the object and write the dimensions in each indicated scale using the guidelines provided. Reference 3.6.

3.5 SI Scales. Measure the object and write the dimensions in each indicated scale using the guidelines provided. Reference 3.6.

3.6 Lines. Draw or sketch the Angle Polygon using the given dimensions. Do not dimension. Reference 3.5.

3.7 Lines, Circles, and Arcs. Draw or sketch the Open Support using the given dimensions. Do not dimension. References 3.5, 3.7, and 3.8.

3.8 Irregular Curves. Plot the points at the intersections shown on the grid provided; then draw a smooth curve through each point using irregular curves. Reference 3.5.6.

Sketching and Text, *Reference Chapter 4*

4.1 Vertical Gothic Lettering. Using HB lead, letter the text in the grided space provided. Use the stroke sequence recommended but only write the letter and not the numbers and arrows shown with each letter and number. All text must be drawn sharp and black. Reference 4.9.

4.2 Inclined Gothic Lettering. Using HB lead, letter the text in the grided space provided. Use the stroke sequence recommended but only write the letter and not the numbers and arrows shown with each letter and number. All text must be drawn sharp and black. Reference 4.9.

4.3 Vertical Gothic Lettering Exercise. Using HB lead, letter the text using the guidelines provided. All text must be drawn sharp and black. Reference 4.9.

4.4 Inclined Gothic Lettering Exercise. Using HB lead, letter the text using the guidelines provided. All text must be drawn sharp and black. Reference 4.9.

4.5 Alphabet of lines. Identify each line type with vertical gothic lettering using the guidelines provided. All text must be drawn sharp and black. Reference 3.4.

4.6 Multiview Sketching 1. Sketch the front, top and right side views using the grided space. Reference 4.7.

4.7 Multiview Sketching 2. Sketch the front, top and right side views using the grided space. Reference 4.7.

4.8 Multiview Sketching 3. Sketch the front, top and right side views using the grided space. Reference 4.7.

4.9 Isometric Sketching 1. Sketch an isometric view of the object shown as multiviews using the grided space. Reference 4.5.

4.10 Isometric Sketching 2. Sketch an isometric view of the object shown as multiviews using the grided space. Reference 4.5.

4.11 Isometric Sketching 3. Sketch an isometric view of the object shown as multiviews using the grided space. Reference 4.5.

4.12 Perspective Sketching. Create a one-point isometric sketch of the object shown. Reference 4.8.

Visualization for Design, *Reference Chapter 5*

5.1 Object Rotation. Target shapes P and Q are to be matched with the correct rotated three-dimensional representations of the letters P and Q. Write the letter P or Q in the square below the rotated letter. Reference 5.5.2.

5.2 Surface Identification 1. In the table, match the given surface letter from the pictorial drawing with the corresponding surface number from the multiview drawing for each view. Reference 5.6.2.

5.3 Surface Identification 2. In the table, match the given surface letter from the pictorial drawing with the corresponding surface number from the multiview drawing for each view. Reference 5.6.2.

5.4 Surface Identification 3. In the table, match the given surface letter from the pictorial drawing with the corresponding surface number from the multiview drawing for each view. Reference 5.6.2.

5.5 Visualization 1. The development (unfolded) is to be matched to one of the five pictorial drawings by circling the correct answer. The development shows the inside surfaces of a three-dimensional object with the shaded portion being the bottom surface.

5.6 Rotation. The object in the top line is rotated to a new position. A new object is to be rotated exactly the same as the first object. Match the correct one of five figures rotated to the corresponding rotation as the rotation is for the first object by circling the correct view. Reference 5.5.2.

5.7 Object Feature Identification. Identify the feature on the object as either an edge (E), face (F), vertex (V), or limiting element (L) in the space provided. Reference 5.4.

5.8 Rotation 2. Which two drawings of the four on the right show the same object as the one on the left? Circle the correct views. Reference 5.5.2.

5.9 Visualization 2. Each problem consists of an object that has been cut by a plane. You must visualize then sketch what the shape of the surface would be if cut by the plane.

5.10 Rotation 3. Each problem consists of a square piece of paper that is folded a number of times before a hole is drilled through it. You must visualize then sketch the unfolded square piece of paper with the resulting holes. Reference 5.5.2.

5.11 Space Relations. If the two pieces on the right were fitted together, what would the resulting figure look like? You must visualize then select from the possible answers on the left what the shape would look like after being fitted together.

Engineering Geometry and Construction, *Reference Chapter 6*

6.1 Coordinates. Four grid lines equal to one unit. In the upper half of the rectangular grid paper, sketch the figure using the following absolute coordinate values; 0,0 3,0 3,2 0,2 and 0,0. In the lower half of the rectangular grid paper, sketch the figure using the following relative coordinate values; 0,0 4,0 0,3 -4,0 and 0, -3. Reference 6.3.

6.2 Coordinates 2. Four grid lines equal to one unit. Using the isometric grid paper and following the right-hand rule, place and label points at the following locations; 0,0,0 4,0,0 4,2,0 0,2,0 0,0,2 4,0,2 4,2,2 0,2,2. After placing the points on the isometric grid connect the following points with lines. Reference 6.3.

1-2, 2-3, 3-4, 4-1.
5-6, 6-7, 7-8, 8-5.
4-8, 3-7, 1-5, 2-6.

6.3 Geometric Construction 1.
   A.   Bisect line AB. Reference Figure 6.28.
   B.   Divide line CD into five equal parts. Reference Figure 6.26.
   C.   Proportionally divide line EF it into three parts. Reference Figure 6.27.
   D.   Draw a line perpendicular to GH through a point 0.5" from one end. Reference Figure 3.51.
   E.   Draw a line parallel to JK and 0.5" from it.
   F.   Draw a line tangent to the given circle. Reference Figure 3.49.

6.4 Geometric Construction 2.
   A.   Construct a line tangent to the arc. Locate and label the point of tangency. Reference Figure 6.35 A.
   B.   Construct a 1" diameter circle tangent to the given circle. Locate and label the point of tangency. Reference Figure 6.35 B.
   C.   Draw a 1" radius arc tangent to the given line. Reference Figure 6.37 A.
   D.   Draw a 0.5" arc tangent to the given lines. Reference Figure 6.37 B.

    E.   Construct a 0.5" radius arctangent to the circle and the line. Reference Figure 6.38.

    F.   Construct a 1" radius arc tangent to the two circles.

6.5 Geometric Construction 3.
    A.   Draw a circle through points A, B, and C. Reference Figure 6.45.
    B.   Draw an ogee curve between the two given lines. Reference Figure 6.48.
    C.   Rectify the given arc. Reference Figure 6.51.
    D.   Construct a parabola using the tangent method. Reference 6.54.
    E.   Construct a hyperbola using the equilateral method. Reference 6.61.
    F.   Locate two foci by constructing a short perpendicular line 0.5" from each end of the given horizontal line. Construct an ellipse using any technique. Reference Figures 6.68 – 6.71.

6.6 Geometric Construction 4.
    A.   Construct a spiral of Archimedes. Reference Figure 6.76.
    B.   Construct an involute of the give circle. Reference Figure 6.82.
    C.   Bisect the given angle. Reference Figure 6.95.
    D.   Transfer the given angle. Reference Figure 6.96.
    E.   Construct a square given one of its sides. Reference Figure 6.103.
    F.   Circumscribe a circle around the given triangle.

6.7 Geometric Construction 5.
    A.   Given the circle, inscribe a square. Reference Figure 6.105.
    B.   Given the circle, circumscribe a square. Reference Figure 6.106.
    C.   Given sides A, B, and C, construct a triangle. Reference Figure 6.108.
    D.   Using the given line construct an equilateral triangle. Reference Figure 6.109.
    E.   Given the circle, inscribe a pentagon. Reference Figure 6.111.
    F.   Given the circle, circumscribe a hexagon. Reference 6.112.

6.8 Geometric Construction 6.
    A.   Given the circle, inscribe a hexagon. Reference Figure 6.113.
    B.   Given the circle, circumscribe an octagon. Reference Figure 6.114.
    C.   Given the circle, inscribe an octagon. Reference Figure 6.115.
    D.   Construct a cycloid. Reference Figure 6.78.
    E.   Construct an epicycloid. Reference Figure 6.79.
    F.   Construct a hypocycloid. Reference Figure 6.79.

6.9 Applied Geometric Construction. Construct the diagram, then measure the X distances. Describe the relationship of the X dimension to the vertical distances.

6.10 Ridge Gasket. Construct the Ridge Gasket using the given metric dimensions.

6.11 Centering Plate. Construct the Centering Plate using the given dimensions.

6.12 Arched Follower. Construct the Arched Follower using the given dimensions.

6.13 Wing Plate. Construct the Wing Plate using the given metric dimensions.

6.14 Transition. Construct the Transition using the given dimensions.

6.15 Variable Guide. Construct the Variable Guide using the given dimensions. Use ½ scale.

Three-Dimensional Modeling, *Reference Chapter 7*

7.1 Circular Sweep 1. Sketch the resulting solid model if the given profiles were to be circularly swept 360 degrees about the Y-axis. Reference 7.6.4.

7.2 Circular Sweep 2. Sketch the resulting solid model if the given profiles were to be circularly swept 360 degrees about the X-axis. Reference 7.6.4.

7.3 Linear Sweep. Sketch the resulting solid model if the given profiles were to be linearly swept 2 units along the +Z-axis. Reference 7.6.4.

7.4 Profile Matching. Twelve objects have been circularly swept from a given profile. Match the swept object with the same profile used to create the object. Reference 7.6.4.

7.5 Boolean Operations 1. Given the three overlapping solid primitives, make an isometric sketch of the resulting solid after applying the following Boolean operations. Reference 7.4.2.
    A.   (A U B)U C
    B.   (A U B) –C
    C.   (A – B) –C

7.6 Boolean Operations 2. Given the three overlapping solid primitives, make an isometric sketch of the resulting solid after applying the following Boolean operations. Reference 7.4.2.
   A. (A – B) –C
   B. (A U B) UC
   C. B- (A U C)

7.7 Boolean Operations 3. Given the three overlapping solid primitives, make an isometric sketch of the resulting solid after applying the following Boolean operations. Reference 7.4.2.
   A. (C – A) –B
   B. (A U C) –B
   C. (A intersect C) –B

7.8 Boolean Operations 4. Given the three overlapping solid primitives, make an isometric sketch of the resulting solid after applying the following Boolean operations; A – B –C. Reference 7.4.2.

7.9 Boolean Operations 5. Given the three overlapping solid primitives, make an isometric sketch of the resulting solid after applying the following Boolean operations; A + B –C. Reference 7.4.2.

7.10 Boolean Operations 6. Given the three overlapping solid primitives, make an isometric sketch of the resulting solid after applying the following Boolean operations; A + B +C. Reference 7.4.2.

7.11 Boolean Operations 7. Given the three overlapping solid primitives, make an isometric sketch of the resulting solid after applying the following Boolean operations; B – C – A. Reference 7.4.2.

7.12 Boolean Operations 8. Given the three overlapping solid primitives, make an isometric sketch of the resulting solid after applying the following Boolean operations; A + C – B. Reference 7.4.2.

7.13 Feature-based Modeling 1. Make a proportional isometric sketch of the resulting solid after applying the feature-based modeling operations. Reference 7.8.3.

7.14 Feature-based Modeling 2. Make a proportional isometric sketch of the resulting solid after applying the feature-based modeling operations. Reference 7.8.3.

7.15 Feature-based Modeling 3. Make a proportional isometric sketch of the resulting solid after applying the feature-based modeling operations. Reference 7.8.3.

Multiview Drawings, *Reference Chapter 8*

8.1 Surface Labeling 1. Draw or sketch the front, top, and right side views of the object shown. Number each visible surface of the multiviews to correspond to the numbers shown in the given view. Reference 8.8.5.
8.2 Surface Labeling 2. Draw or sketch the front, top, and right side views of the object shown. Number each visible surface in of the multiviews to correspond to the numbers shown in the given view. Identify each surface as normal (N), inclined (I) or oblique (O) using the table provided. Reference 8.8.5.
8.3 Missing Views 1. Given the two views of a multiview drawing, sketch or draw the missing view. Reference 8.8.
8.4 Missing Views 2. Given the two views of a multiview drawing, sketch or draw the missing view. Reference 8.8.
8.5 Missing Lines 1. Given three incomplete views of a multiview drawing of an object, sketch or draw missing line or lines. Reference 8.8.
8.6 Missing Lines 2. Given three incomplete views of a multiview drawing of an object, sketch or draw missing line or lines. Reference 8.8.
8.7 Multiview Drawing 1. Given the isometric view of an object, sketch or draw three views. Reference 8.8.
8.8 Multiview Drawing 2. Given the isometric view of an object, sketch or draw three views. Reference 8.8.

8.9 Multiview Drawing 2. Given the isometric view of an object, sketch or draw three views. Reference 8.8.

8.10 Line Identification. Identify labeled lines as either normal (N), inclined (I), or oblique (O). Reference 8.6.1.

8.11 View Identification 1. Select the correct view indicated by the arrow. Reference 8.8.

8.12 View Identification 2. Select the correct right side view for the given top and front views. Reference 8.8.

8.13 Snubber. Sketch or draw multiviews of the object shown in the pictorial. Reference 8.4.

8.14 Spacer. Sketch or draw multiviews of the object shown in the pictorial. Reference 8.4.

8.15 Motor Plate. Sketch or draw multiviews of the object shown in the pictorial. Reference 8.4.

8.16 Speed Spacer. Sketch or draw multiviews of the object shown in the pictorial. Reference 8.4.

8.17 Shaft Support. Sketch or draw multiviews of the object shown in the pictorial. Reference 8.4.

8.18 Gear Index. Sketch or draw multiviews of the object shown in the pictorial. Reference 8.4.

8.19 Stop Base. Sketch or draw multiviews of the object shown in the pictorial. Reference 8.4.

8.20 Locating Block. Sketch or draw multiviews of the object shown in the pictorial. Reference 8.4.

8.21 Cover Guide. Sketch or draw multiviews of the object shown in the pictorial. Reference 8.4.

8.22 Dial Bracket. Sketch or draw multiviews of the object shown in the pictorial. Reference 8.4.

8.23 Bearing Block. Sketch or draw multiviews of the object shown in the pictorial. Reference 8.4.

8.24 Pulley Support. Sketch or draw multiviews of the object shown in the pictorial. Reference 8.4.

8.25 Centering Clip. Sketch or draw multiviews of the object shown in the pictorial. Reference 8.4.

8.26 Adjustable Guide. Sketch or draw multiviews of the object shown in the pictorial. Reference 8.4.

8.27 Bar Hinge. Sketch or draw multiviews of the object shown in the pictorial. Reference 8.4.

8.28 Dryer Clip. Sketch or draw multiviews of the object shown in the pictorial. Reference 8.4.

8.29 Retaining Cap. Sketch or draw multiviews of the object shown in the pictorial. Reference 8.4.

8.30 Locating Base. Sketch or draw multiviews of the object shown in the pictorial. Reference 8.4.

8.31 Anchor Base. Sketch or draw multiviews of the object shown in the pictorial. Reference 8.4.

8.32 Strike Arm. Sketch or draw multiviews of the object shown in the pictorial. Reference 8.4.

8.33 Retainer Clip. Sketch or draw multiviews of the object shown in the pictorial. Reference 8.4.

8.34 Bearing Plate. Sketch or draw multiviews of the object shown in the pictorial. Reference 8.4.

8.35 Drive Base. Sketch or draw multiviews of the object shown in the pictorial. Reference 8.4.

8.36 Cutoff. Sketch or draw multiviews of the object shown in the pictorial. Reference 8.4.

Axonometric and Oblique Drawings, *Reference Chapter 9*

9.1 Isometric Drawing 1. Given the orthographic views, sketch or draw the object in isometric using the isometric grid. Double the object size on the isometric grid. Reference 9.2.1.

9.2 Isometric Drawing 2. Given the orthographic views, sketch or draw the object in isometric using the isometric grid. Double the object size on the isometric grid. Reference 9.2.1.

9.3 Isometric Drawing 3. Given the orthographic views, sketch or draw the object in isometric using the isometric grid. Double the object size on the isometric grid. Reference 9.2.1.

9.4 Isometric Drawing 4. Given the orthographic views, sketch or draw the object in isometric using the isometric grid. Double the object size on the isometric grid. Reference 9.2.1.

9.5 Isometric Drawing 5. Given the orthographic views, sketch or draw the object in isometric using the isometric grid. Double the object size on the isometric grid. Reference 9.2.1.

9.6 Isometric Drawing 6. Given the orthographic views, sketch or draw the object in isometric using the isometric grid. Double the object size on the isometric grid. Reference 9.2.1.

9.7 Isometric Drawing 7. Given the orthographic views, sketch or draw the object in isometric using the isometric grid. Double the object size on the isometric grid. Reference 9.2.1.

9.8 Oblique Drawing 1. Given the isometric view, sketch or draw the object as a cabinet oblique using the grid. Double the object size on the rectangular grid. Reference 9.5.

9.9 Oblique Drawing 2. Given the isometric view, sketch or draw the object as a cabinet oblique using the grid. Double the object size on the rectangular grid. Reference 9.5.

9.10 Oblique Drawing 3. Given the isometric view, sketch or draw the object as a cabinet oblique using the grid. Double the object size on the rectangular grid. Reference 9.5.

9.11 Oblique Drawing 4. Given the isometric view, sketch or draw the object as a cabinet oblique using the grid. Double the object size on the rectangular grid. Reference 9.5.

9.12 Oblique Drawing 5. Given the isometric view, sketch or draw the object as a cabinet oblique using the grid. Double the object size on the rectangular grid. Reference 9.5.

9.13 Oblique Drawing 6. Given the isometric view and each grid equal to 0.5", sketch or draw the object as a cabinet oblique. Double the object size on the rectangular grid. Reference 9.5.

9.14 Oblique Drawing 7. Given the isometric view and each grid equal to 0.5", sketch or draw the object as a cabinet oblique. Double the object size on the rectangular grid. Reference 9.5.

9.15 Oblique Drawing 8. Given the isometric view and each grid equal to 0.5", sketch or draw the object as a cabinet oblique. Double the object size on the rectangular grid. Reference 9.5.

9.16 Isometric Drawing 7. Given the multiviews and each grid equal to 0.5", sketch or draw the object as an isometric view. Double the object size on the isometric grid. Reference 9.2.1.

9.17 Isometric Drawing 8. Given the multiviews and each grid equal to 0.5", sketch or draw the object as an isometric view. Double the object size on the isometric grid. Reference 9.2.1.

9.18 Isometric Drawing 9. Given the multiviews and each grid equal to 0.5", sketch or draw the object as an isometric view. Double the object size on the isometric grid. Reference 9.2.1.

9.19 Isometric Drawing 10. Given the multiviews and each grid equal to 0.5", sketch or draw the object as an isometric view. Double the object size on the isometric grid. Reference 9.2.1.

9.20 Isometric Drawing 11. Given the multiviews and each grid equal to 0.5", sketch or draw the object as an isometric view. Double the object size on the isometric grid. Reference 9.2.1.

Perspective Drawings, *Reference Chapter 10*

10.1 Perspective Identification. Identify the important parts of this perspective drawing. Reference 10.2.

10.2 One-point Perspective 1. Create a one-point perspective drawing of the object using the given reference points. Reference 10.5.

10.3 One-point Perspective 2. Create a one-point perspective drawing of the object using the given reference points. Reference 10.5.

10.4 One-point Perspective 3. Create a one-point perspective drawing of the object using the given reference points. Reference 10.5.

10.5 One-point Perspective 4. Create a one-point perspective drawing of the object using the given reference points. Reference 10.5.

10.6 Two-point Perspective 1. Create a two-point perspective drawing of the object using the given reference points. Reference 10.6.

10.7 Two-point Perspective 2. Create a two-point perspective drawing of the object using the given reference points. Reference 10.6.

10.8 Two-point Perspective 3. Create a two-point perspective drawing of the object using the given reference points. Reference 10.6.

Auxiliary Views, *Reference Chapter 11*

11.1 Auxiliary View 1. Draw the two given views then create a partial auxiliary view of the inclined surface. Each grid space equals 0.5". Reference 11.1.

11.2 Auxiliary View 2. Draw the two given views then create a partial auxiliary view of the inclined surface. Each grid space equals 0.5". Reference 11.1.

11.3 Auxiliary View 3. Draw the two given views then create a partial auxiliary view of the inclined surface. Each grid space equals 0.5". Reference 11.1.

11.4 Auxiliary View 4. Draw the two given views then create a partial auxiliary view of the inclined surface. Each grid space equals 0.5". Reference 11.1.

11.5 Auxiliary View 5. Draw the two given views then create a partial auxiliary view of the inclined surface. Each grid space equals 0.5". Reference 11.1.

11.6 Auxiliary View 6. Draw the two given views then create a partial auxiliary view of the inclined surface. Each grid space equals 0.5". Reference 11.1.

11.7 Auxiliary View 7. Draw the two given views then create a partial auxiliary view of the inclined surface. Each grid space equals 0.5". Reference 11.1.

11.8 Auxiliary View 8. Draw the necessary views, including a complete auxiliary view of the inclined or oblique surface. Reference 11.1, 2, and 3.

11.9 Auxiliary View 9. Draw the two given views then create a partial auxiliary view of the inclined surface. Reference 11.1.

11.10 Auxiliary View 10. Draw the two given views then create a partial auxiliary view of the inclined surface. Reference 11.1.

11.11 Auxiliary View 11. Draw the two given views then create a partial auxiliary view of the inclined surface. Reference 11.1.

11.12 Auxiliary View 12. Draw the two given views then create a partial auxiliary view of the inclined surface. Reference 11.1.

11.13 Auxiliary View 13. Draw the two given views then create a partial auxiliary view of the inclined surface. Reference 11.1.

11.14 Auxiliary View 14. Draw the two given views then create a partial auxiliary view of the inclined surface. Reference 11.1.

11.15 Auxiliary View 15. Draw the two given views then create a partial auxiliary view of the inclined surface. Reference 11.1.

11.16 Auxiliary View 16. Draw the two given views then create a partial auxiliary view of the inclined surface. Reference 11.1.

11.17 Auxiliary View 17. Draw the two given views then create a partial auxiliary view of the inclined surface. Reference 11.1.

11.18 Auxiliary View 18. Draw the two given views then create a partial auxiliary view of the inclined surface. Reference 11.1.

11.19 Auxiliary View 19. Draw the two given views then create a partial auxiliary view of the inclined surface. Reference 11.1.

11.20 Auxiliary View 20. Draw the two given views then create a partial auxiliary view of the inclined surface. Reference 11.1.

11.21 Auxiliary View 21. Draw the two given views then create a partial auxiliary view of the inclined surface. Reference 11.1.

11.22 Auxiliary View 22. Draw the two given views then create a partial auxiliary view of the inclined surface. Reference 11.1.

Fundamentals of Descriptive Geometry, *Reference Chapter 12*

12.1 Line Identification. Name the types of lines in the space provided. Reference 12.5.

12.2 Point Identification. Answer the following questions in the space provided. Reference 12.5.
  A. Name the type of line for LM.
  B. Which point(s) is/are closet to the viewer when looking at the front view?
  C. Which point(s) is/are farthest away from the viewer when looking at the front view?
  D. Which point(s) is/are highest to the viewer when looking at the front view?
  E. Which point(s) is/are lowest to the viewer when looking at the front view?

12.3 True Length Lines 1. Reference 12.5.3.
  A. Draw the true-length view of line AB.
  B. Draw the true-length view of line CD.
  C. Draw the true-length view of line EG, and measure its length.
  D. Complete the front view, and find the true-length view of line JK.

12.4 True Length Lines 2. Reference 12.5.3.
  A. The true length of line AB is 1.75" or 44 mm. Point A is in front of point B. Complete the top view.
  B. The true length of line CD is 2.50" or 64 mm. Point D is in front of point C. Complete the top view.
  C. The true length of line ED is 1.75" or 44 mm. Point E is below point D. Complete the front and profile views.
  D. The true length of line JK is 1.50" or 38 mm. Point J is above point K. Complete the front view.

12.5 True Length Lines 3. Reference 12.5.3.
  A. Draw the true-length view of line AB.
  B. Draw the true-length view of line CD.
  C. Find the true length of line EG.
  D. Complete the front view, and find the true length of line JK.

12.6 True Length Lines 4. Find the true-lengths of lines A1, B2, C3, and D4. Reference 12.5.3.

12.7 Surface Area. Find the lateral surface area of the pyramid. Reference 12.6.

12.8 Edge View of Planes 1. Points A, B, and C form a plane in space. Show an edge view of the plane. Reference 12.6.2.

12.9 Edge View of Planes 2. Points D, E, and F form a plane in space. Show an edge view of the plane. Reference 12.6.2.

12.10 Edge View of Planes 3. Given points J, K, L, and M, draw an edge view of the plane formed by the points. Reference 12.6.2.

12.11 True Size of Planes 1. Draw the true size of planes AOC and DLN. Reference 12.6.3.

12.12 True Size of Planes 2. Find the area of the largest inscribed circle in the oblique plane MNO. Reference 12.6.3.

12.13 Angle Between Planes. Find the dihedral angle between planes LGM and LGK. The object is a tetrahedron. Reference 12.6.4.

Intersections and Developments, *Reference Chapter 13*

13.1 Development 1. Create a development of the object shown. Reference 13.3.

13.2 Development 2. Create a development of the object shown. Reference 13.3.

13.3 Development 3. Create a development of the object shown. Reference 13.3.

13.4 Development 4. Create a development of the object shown. Reference 13.3.

13.5 Development 5. Create a development of the object shown. Reference 13.3.

13.6 Development 6. Create a development of the object shown. Reference 13.3.

13.7 Development 7. Create a development of the object shown. Reference 13.3.

13.8 Development 8. Create a development of the object shown. Reference 13.3.

13.9 Development 9. Create a development of the object shown. Reference 13.3.

13.10 Development 10. Create a development of the object shown. Reference 13.3.

13.11 Development 11. Create a development of the object shown. Reference 13.3.

13.12 Intersections 1. Create the line of intersection between the two objects. Reference 13.2.

13.13 Intersections 2. Create the line of intersection between the two objects. Reference 13.2.

13.14 Intersections 3. Create the line of intersection between the two objects. Reference 13.2.

13.15 Intersecting Planes 1. Create the line of intersection between the planes. Reference 13.2.

13.16 Intersecting Planes 2. Create the line of intersection between the planes. Reference 13.2.

13.17 Intersecting Lines and Planes. Create the intersecting point between the line and the cylinder. Reference 13.2.

Section Views, *Reference Chapter 14*

14.1 Section Views 1. Sketch or draw the section view as indicated by the cutting plane line. Reference 14. 4.

14.2 Section Views 2. Sketch or draw the section view as indicated by the cutting plane line. Reference 14. 4.

14.3 Counter Block. Draw the necessary views including a full section view of the object. Reference 14.4.

14.4 Bracket. Draw the necessary views including an offset section view of the object. Reference 14.4.

14.5 Ring Collar. Draw the necessary views including a half section view of the object. Reference 14.4.

14.6 Axle Center. Draw the necessary views including a full section view of the object. Reference 14.4.

14.7 Taper Collar. Draw the necessary views including a half section view of the object. Reference 14.4.

14.8 Section View Matching. Using the space provided, identify the correct section view given the front and top view and the location of the cutting plane. Reference 14.4.

Dimensioning and Tolerancing Practices, *Reference Chapter 15*

15.1 Dimensioning 1. Use sketching to fully dimension the objects. Reference 15.3.

15.2 Dimensioning 2. Use sketching to fully dimension the objects. Reference 15.3.

15.3 Angle Bracket. Draw three views of the object then dimension. Reference 15.3.

15.4 Angle Clamp. Draw three views of the object then dimension. Reference 15.3.

15.5 Offset Strap. Draw three views of the object then dimension. Reference 15.3.

15.6 Mill Fixture Base. Draw three views of the object then dimension. Reference 15.3.

15.7 Tolerancing 1. Use the tolerancing tables at the back of your textbook to calculate the limit dimensions between the shaft and the hole for the given class of fit. Reference 15.6.12.

15.8 Tolerancing 2. Determine the limit dimensions between the shoulder screw and bushing and between the bushing and the housing using the specified fit tolerancing tables found in the back of your textbook. Reference 15.6.12.

15.9 Dimension the features indicated on the given drawing. Reference

Geometric Dimensioning and Tolerancing Basics, *Reference Chapter 16*

16.1 Geometric Tolerancing 1. Sketch the geometric dimensioning and tolerancing symbol in the space provided. Reference 16.2.

16.2 Geometric Tolerancing Tolerancing 2. Dimension the drawings including feature control frames. Reference 16.2.

16.3 Geometric Tolerancing 3. Dimension the drawings including feature control frames. Reference 16.2.

Fastening Devices and Methods, *Reference Chapter 17*

17.1. Drawing Fasteners. In the space provided draw the side and top views of the following fasteners. Reference 17.8.
   A. 1 x 2 ¼ hex head cap screw, 1-8NC-3.
   B. 1 x 2 ¼ flat head cap screw, 1-8NC-3.
   C. 1 x 1 ¾ fillet head cap screw, 1-8NC-3.
   D. 1 x 2 ¼ button head cap screw, 1-8NC-3.
   E. 1 x 2 ¼ socket head cap screw, 1-8NC-3.

17.2 Standard Fasteners. In the space provided, draw the side and top views of the following standard fasteners. Reference 17.8.
   A. 1 x 2.25 hex head bolt
   B. 1 x 2.25 stud
   C. 1 x 2.25 square head bolt
   D. 0.5 x 1.00 square head set screw
   E. 0.75 x 1.50 hexagon set screw, flat point

17.3 Graphic Thread Representation. In the space provided, draw the following; Reference 17.6
   A. End and side schematic view of a 0.75 tapped through hole.
   B. End and side schematic section view of a 0.75 tapped through hole.
   C. End and side schematic view of a 0.75 tapped 1.5" blind hole.
   D. End and side schematic view of a 0.75 x 1.5" bottomed tapped hole.
   E. End and side schematic section view of a 0.75 tapped through hole.

Working Drawings, *Reference Chapter 19*

19.1 V-Block. Create a complete set of working drawings including a parts list. Reference 19.1.

19.2 Belt Tightener. Create a complete set of working drawings including a parts list. Reference 19.1.

Technical Data Presentation, *Reference Chapter 20*

20.1 Bar Graph. Create a single bar graph of the given data. Reference 20.3.1.

Mechanisms: Gears, Cams, Bearings, and Linkages, *Reference Chapter 22*

22.1 Spur Gear. Using the data found in the table, calculate the dimensions of the spur gear and create a detail drawing and a cutting data table. Reference 22.2.9.

22.2 Cam Profile. Create the displacement diagram and cam profile using the information provided. Reference 22.3.5.

Ten blank sheets
Ten blank square grid sheets
Ten blank isometric grid sheets

203 total problems

C. Draw 8 equally spaced 45-degree lines.

F. Draw 8 equally spaced 75-degree lines.

B. Draw 6 equally spaced vertical lines.

E. Draw 8 equally spaced 15-degree lines.

A. Draw 6 equally spaced horizontal lines.

D. Draw 8 equally spaced 30-degree lines.

| LINE DRAWING 1 | REFERENCE UNIT 3.5 | NAME __ COURSE __ | DATE __ | DRAWING 3.1 |

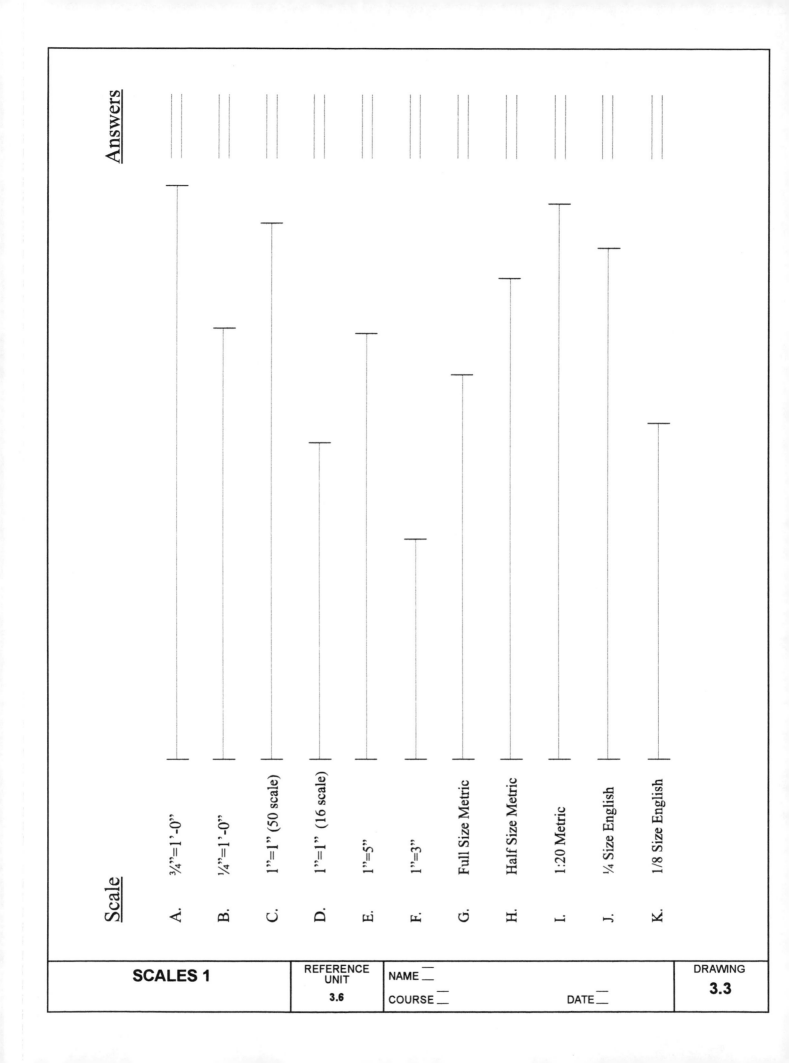

Answers

Scale

A. ¾"=1'-0"

B. ¼"=1'-0"

C. 1"=1" (50 scale)

D. 1"=1" (16 scale)

E. 1"=5"

F. 1"=3"

G. Full Size Metric

H. Half Size Metric

I. 1:20 Metric

J. ¼ Size English

K. 1/8 Size English

| SCALES 1 | REFERENCE UNIT 3.6 | NAME __ COURSE __ | DATE __ | DRAWING 3.3 |

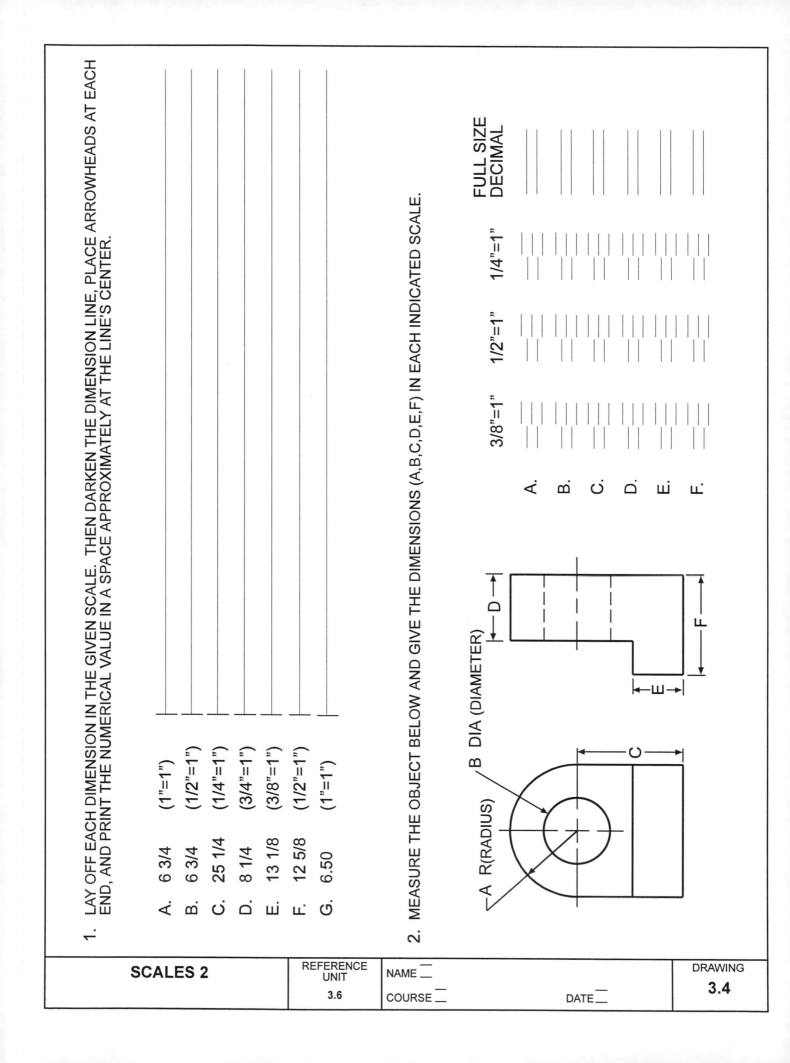

1. LAY OFF EACH DIMENSION IN THE GIVEN SCALE. THEN DARKEN THE DIMENSION LINE, PLACE ARROWHEADS AT EACH END, AND PRINT THE NUMERICAL VALUE IN A SPACE APPROXIMATELY AT THE LINE'S CENTER.

A. 6 3/4   (1"=1")

B. 6 3/4   (1/2"=1")

C. 25 1/4   (1/4"=1")

D. 8 1/4   (3/4"=1")

E. 13 1/8   (3/8"=1")

F. 12 5/8   (1/2"=1")

G. 6.50   (1"=1")

2. MEASURE THE OBJECT BELOW AND GIVE THE DIMENSIONS (A,B,C,D,E,F) IN EACH INDICATED SCALE.

FULL SIZE
DECIMAL

3/8"=1"   1/2"=1"   1/4"=1"

A.
B.
C.
D.
E.
F.

A R(RADIUS)

B DIA (DIAMETER)

| SCALES 2 | REFERENCE UNIT 3.6 | NAME __ COURSE __ DATE __ | DRAWING 3.4 |

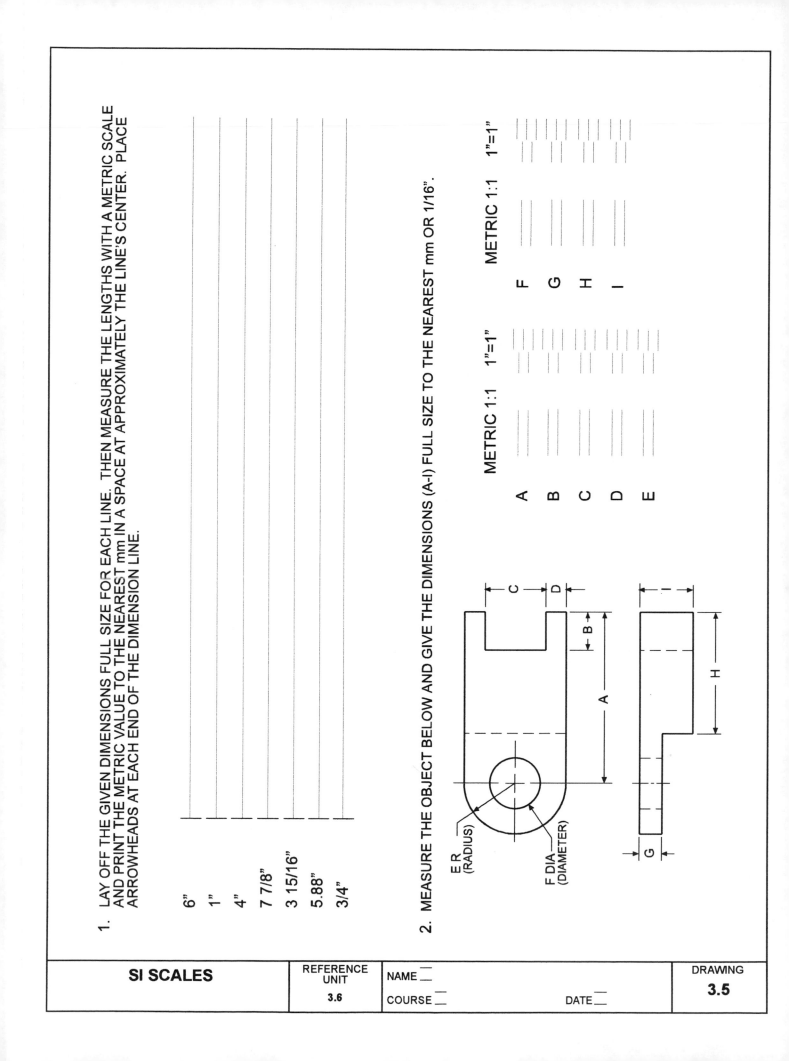

1. LAY OFF THE GIVEN DIMENSIONS FULL SIZE FOR EACH LINE. THEN MEASURE THE LENGTHS WITH A METRIC SCALE AND PRINT THE METRIC VALUE TO THE NEAREST mm IN A SPACE AT APPROXIMATELY THE LINE'S CENTER. PLACE ARROWHEADS AT EACH END OF THE DIMENSION LINE.

6"

1"

4"

7 7/8"

3 15/16"

5.88"

3/4"

2. MEASURE THE OBJECT BELOW AND GIVE THE DIMENSIONS (A-I) FULL SIZE TO THE NEAREST mm OR 1/16".

METRIC 1:1   1"=1"   METRIC 1:1   1"=1"

A   F

B   G

C   H

D   I

E

E R
(RADIUS)

F DIA
(DIAMETER)

| SI SCALES | REFERENCE UNIT 3.6 | NAME __ COURSE __ DATE __ | DRAWING 3.5 |

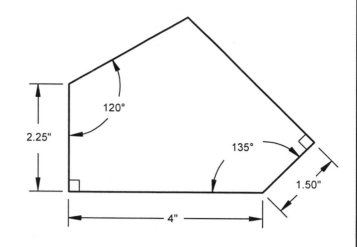

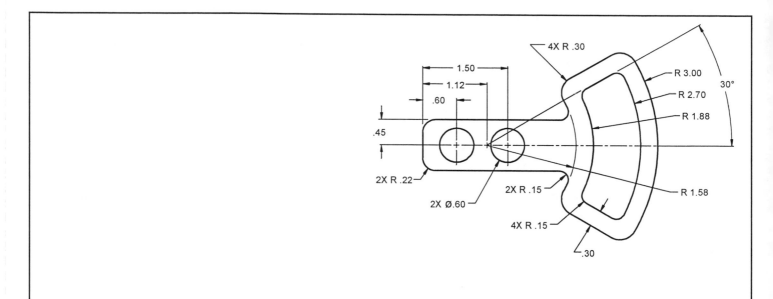

LINES, CIRCLES, AND ARCS	REFERENCE UNIT	NAME __		DRAWING
	3.5, 3.7, 3.8	COURSE __	DATE __	3.7

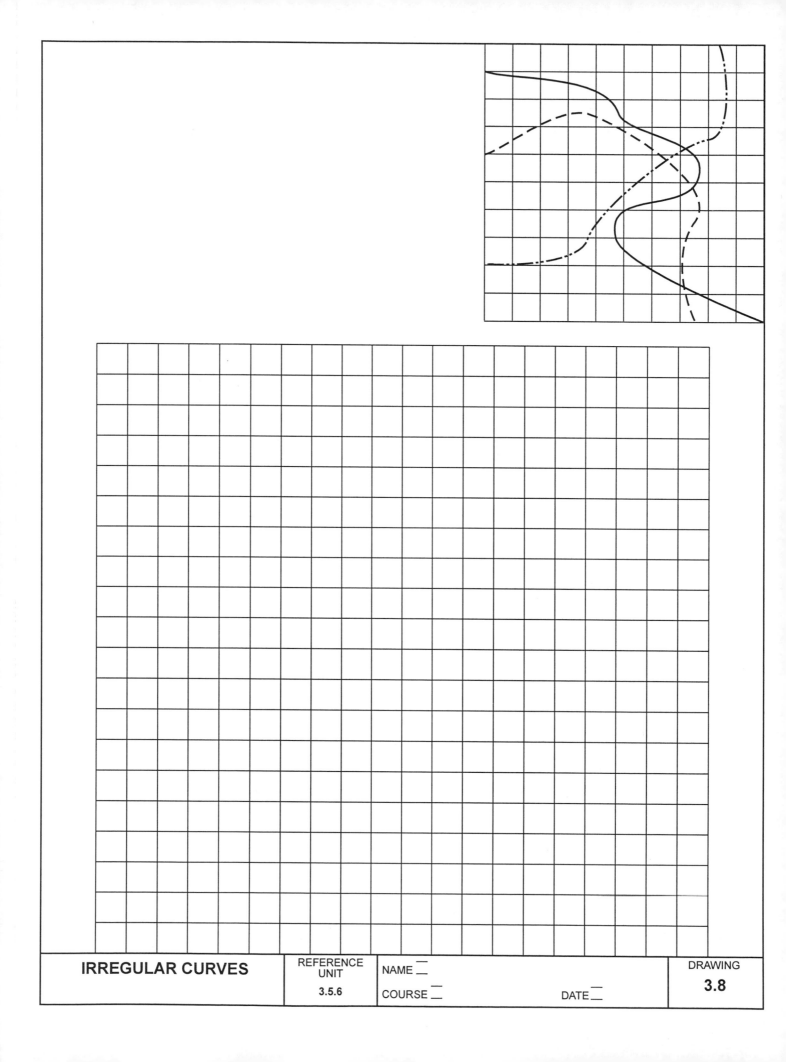

IRREGULAR CURVES

REFERENCE UNIT

3.5.6

NAME __

COURSE __

DATE __

DRAWING

3.8

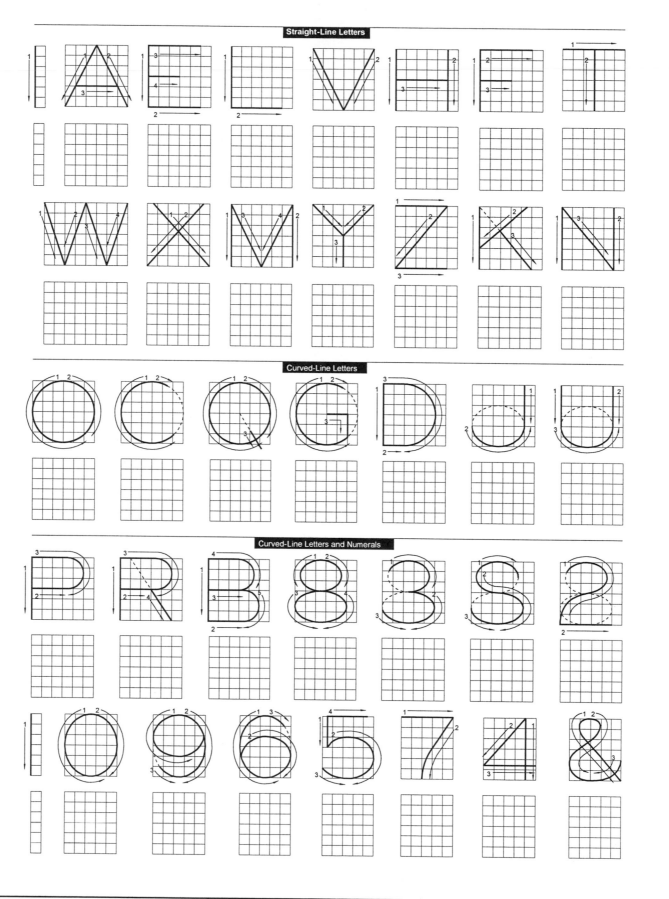

Straight-Line Letters

Curved-Line Letters

Curved-Line Letters and Numerals

| VERTICAL GOTHIC LETTERING | REFERENCE UNIT 4.9 | NAME __ COURSE __ DATE __ | DRAWING 4.1 |

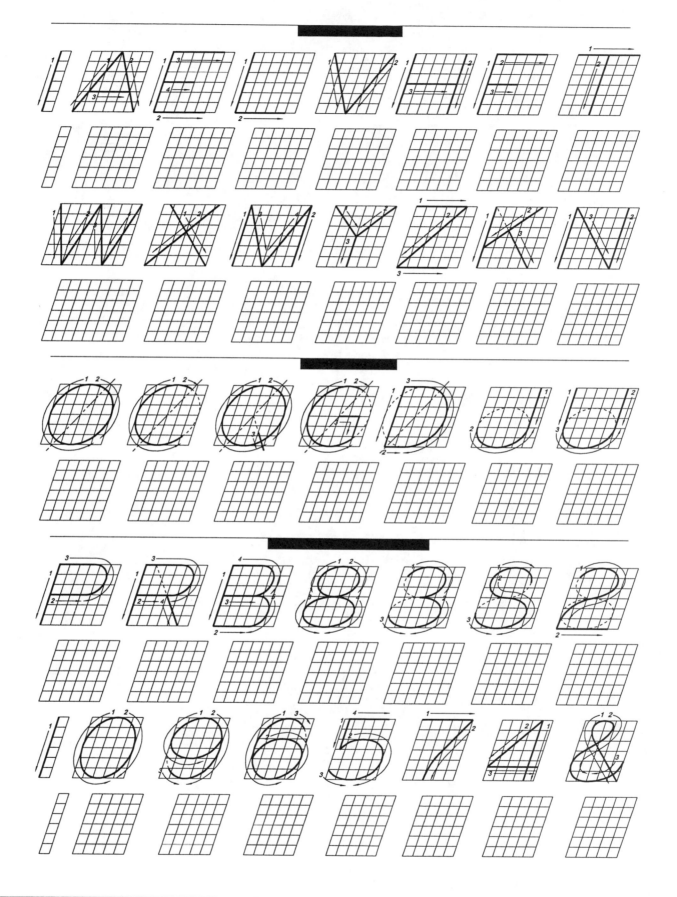

| INCLINED GOTHIC LETTERING | REFERENCE UNIT 4.9 | NAME ___ COURSE ___ DATE ___ | DRAWING 4.2 |

WHILE IT IS TRUE THAT

"PRACTICE MAKES PERFECT," IT

MUST BE UNDERSTOOD THAT

PRACTICE IS NOT ENOUGH, BUT IT

MUST BE ACCOMPANIED BY A CON-

TINUOUS EFFORT TO IMPROVE. EXCEL-

LENT LETTERERS ARE OFTEN NOT GOOD

WRITERS. USE A FAIRLY SOFT PENCIL, AND AL-

WAYS KEEP IT SHARP, ESPECIALLY FOR SMALL

LETTERS. MAKE THE LETTERS CLEAN-CUT AND

DARK-NEVER FUZZY, GRAY, OR INDEFINITE. 1234

1 1/2 1.500 3/16 45'-6 32 ° 15.489 13/64 12"=1'-0 7 5/16 12.3 1/2 2 1/4

ONE MUST HAVE A CLEAR MENTAL IMAGE OF THE LETTERS. 234

| VERTICAL GOTHIC LETTERING EXERCISE | REFERENCE UNIT 4.9 | NAME __ COURSE __ | DATE __ | DRAWING 4.3 |

WHILE IT IS TRUE THAT

"PRACTICE MAKES PERFECT," IT

MUST BE UNDERSTOOD THAT

PRACTICE IS NOT ENOUGH, BUT IT

MUST BE ACCOMPANIED BY A CON-

TINUOUS EFFORT TO IMPROVE. EXCEL-

LENT LETTERERS ARE OFTEN NOT GOOD

WRITERS. USE A FAIRLY SOFT PENCIL, AND AL-

WAYS KEEP IT SHARP, ESPECIALLY FOR SMALL

LETTERS. MAKE THE LETTERS CLEAN-CUT AND

DARK-NEVER FUZZY, GRAY, OR INDEFINITE. 1234

1 1/2 1.500 3/16 45'-6 32 ° 15.489 13/64 12"=1'-0 7 5/16 12.3 1/2 2 1/4

ONE MUST HAVE A CLEAR MENTAL IMAGE OF THE LETTERS. 234

| INCLINED GOTHIC LETTERING EXERCISE | REFERENCE UNIT 4.9 | NAME __ COURSE __ | DATE __ | DRAWING 4.4 |

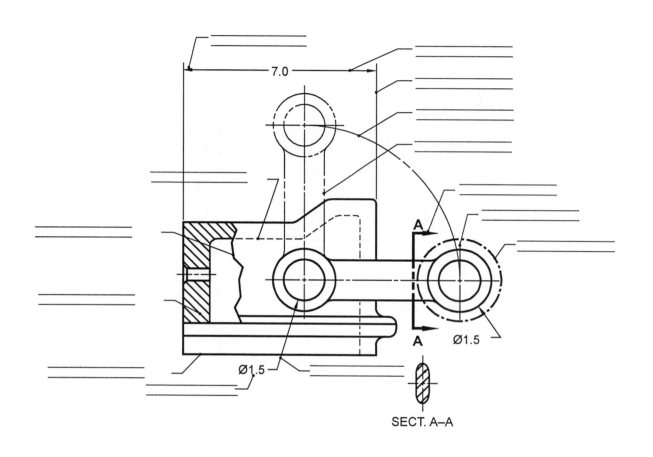

7.0

A

Ø1.5

A

Ø1.5

SECT. A–A

ALPHABET OF LINES

REFERENCE UNIT

3.4

NAME __

COURSE __

DATE __

DRAWING

4.5

A.

B.

C.

D.

A.

B.

C.

D.

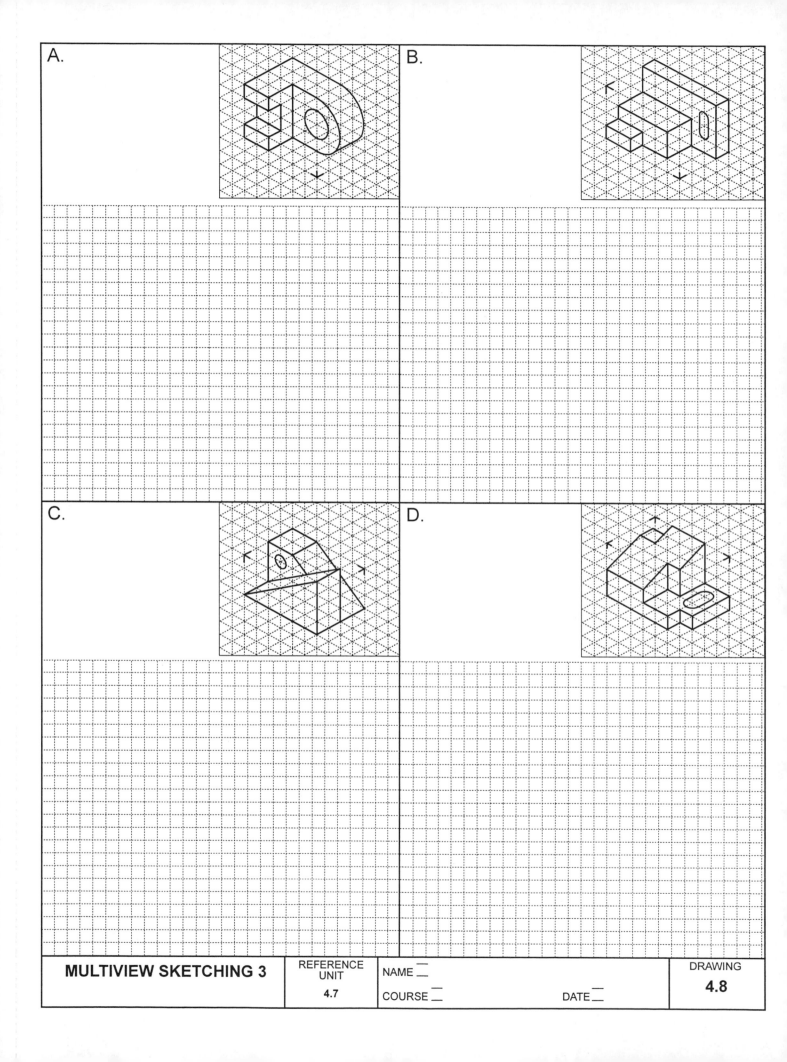

A.

B.

C.

D.

MULTIVIEW SKETCHING 3

REFERENCE UNIT

4.7

NAME __

COURSE __

DATE __

DRAWING

4.8

A.

B.

C.

D.

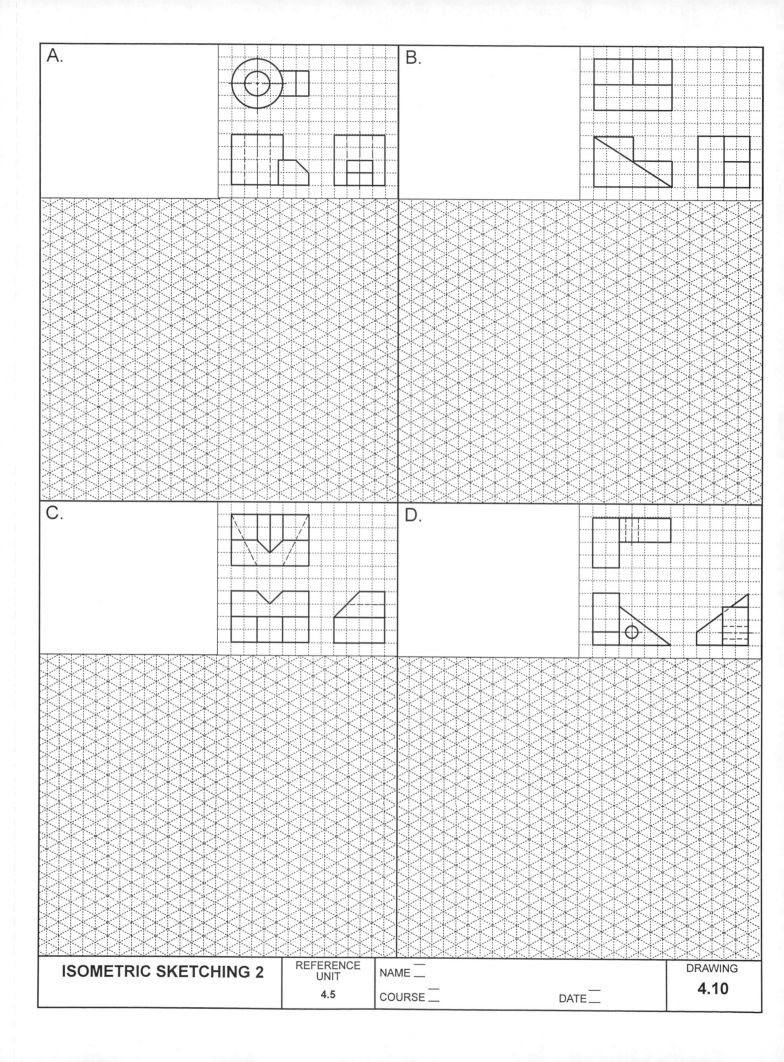

A.

B.

C.

D.

A.

B.

C.

D.

ISOMETRIC SKETCHING 3

REFERENCE
UNIT

4.5

NAME —

COURSE —

DATE —

DRAWING

4.11

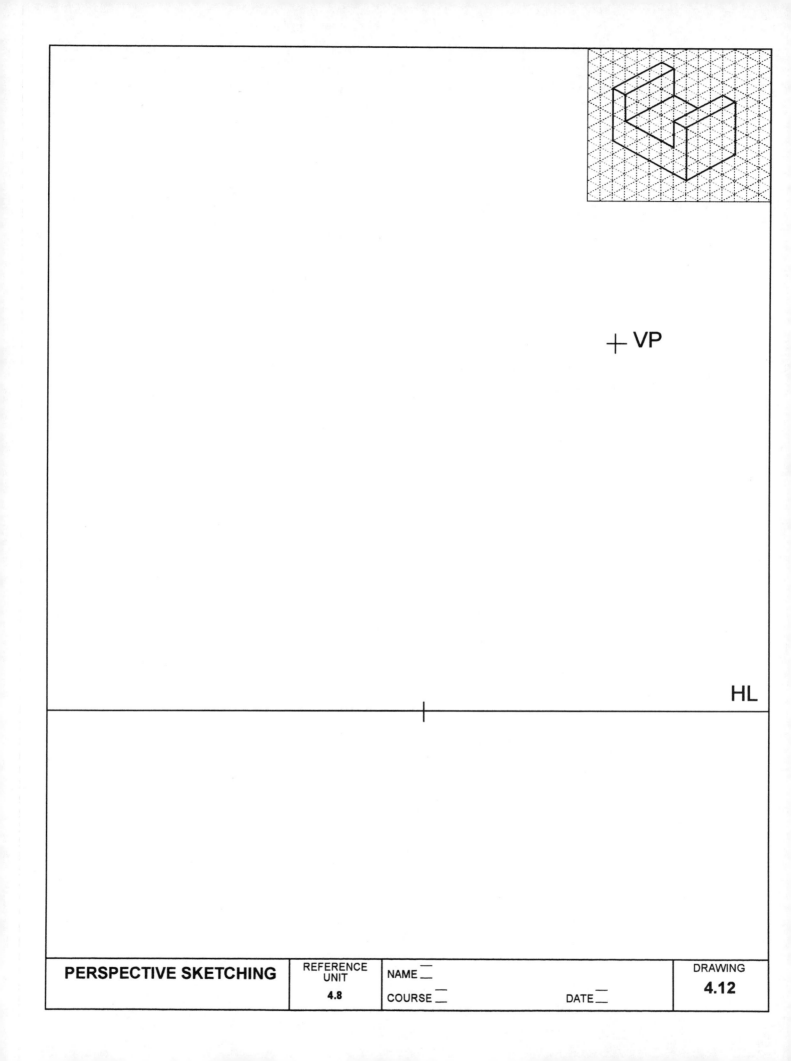

+ VP

HL

TARGET SHAPES

1 ☐ 2 ☐ 3 ☐ 4 ☐ 5 ☐ 6 ☐ 7 ☐ 8 ☐ 9 ☐ 10 ☐ 11 ☐ 12 ☐ 13 ☐ 14 ☐

OBJECT ROTATION

REFERENCE UNIT

5.5.2

NAME __

COURSE __

DATE __

DRAWING

5.1

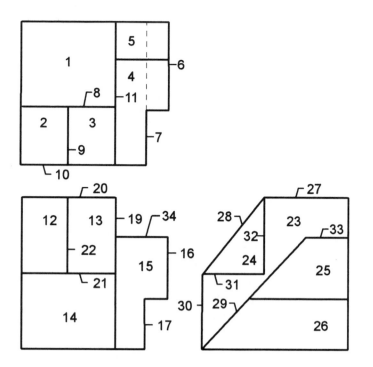

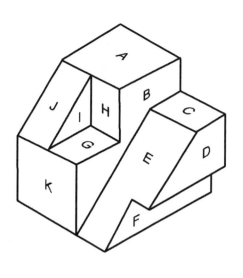

Surface	Top	Front	Side
A			
B			
C			
D			
E			
F			
G			
H			
I			
J			
K			

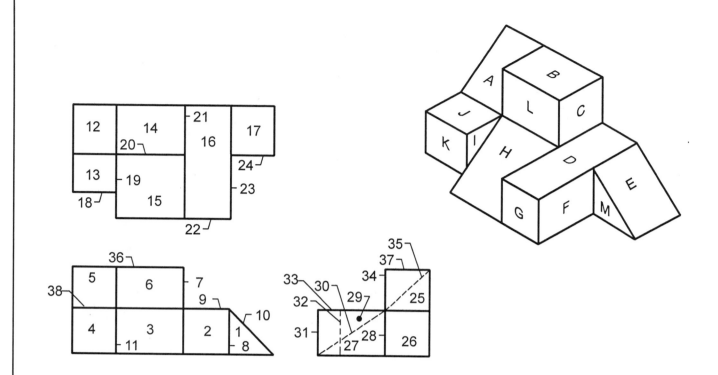

Surface	Top	Front	Side
A			
B			
C			
D			
E			
F			
G			
H			
I			
J			
K			
L			
M			

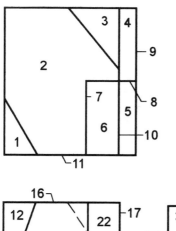

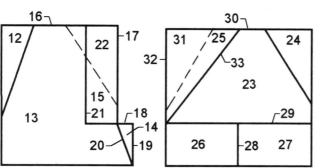

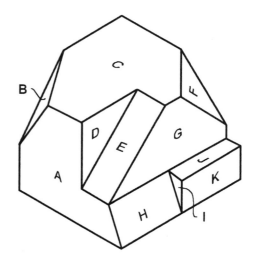

Surface	Top	Front	Side
A			
B			
C			
D			
E			
F			
G			
H			
I			
J			
K			

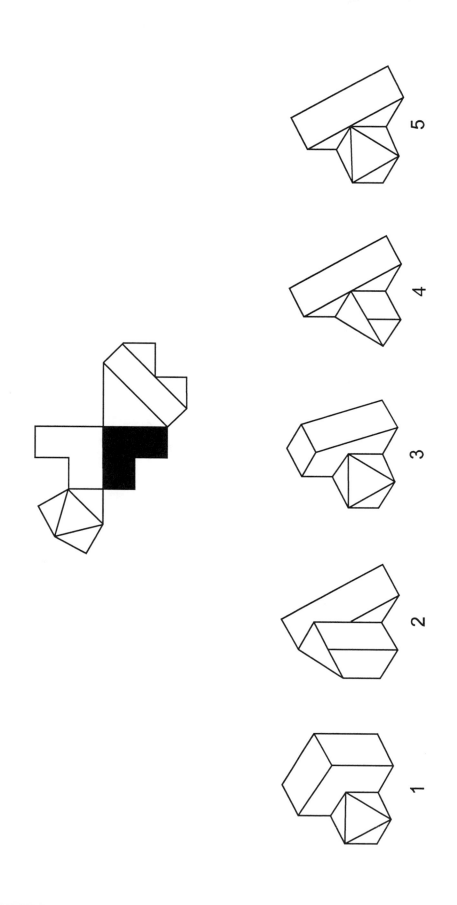

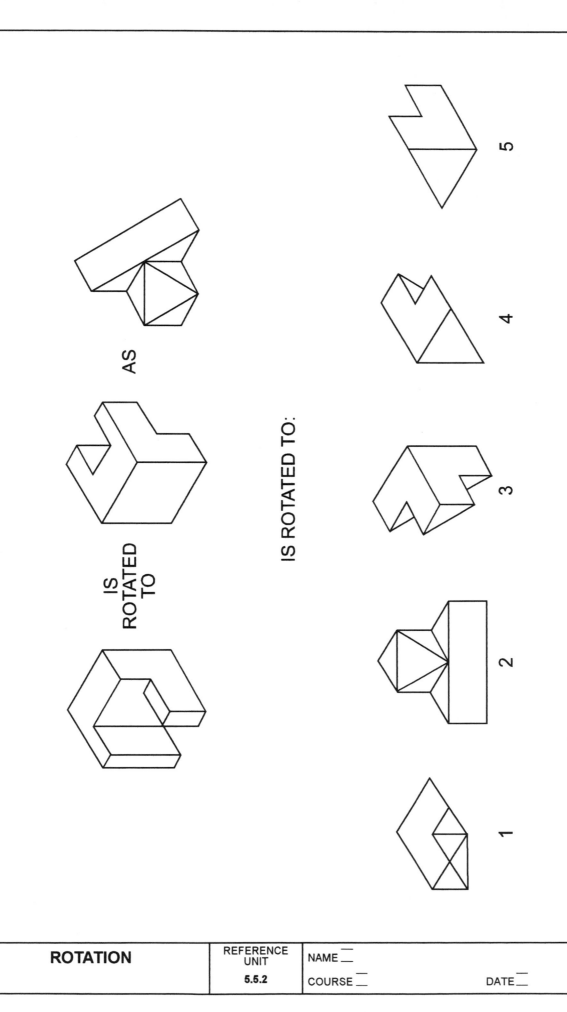

AS

IS
ROTATED
TO

IS ROTATED TO:

1

2

3

4

5

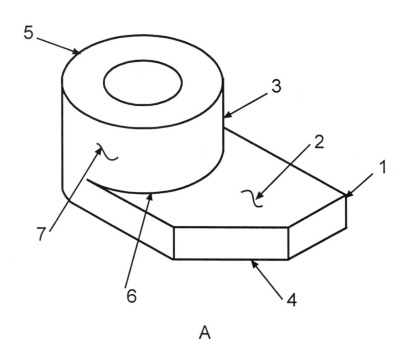

5

3

2

1

7

6

4

A

__A__

1. ____

2. ____

3. ____

4. ____

5. ____

6. ____

7. ____

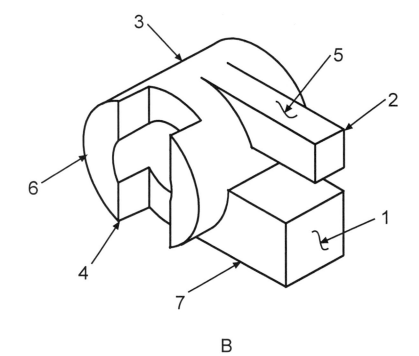

3

5

2

6

4

1

7

B

__B__

1. ____

2. ____

3. ____

4. ____

5. ____

6. ____

7. ____

| OJBECT FEATURE IDENTIFICATION | REFERENCE UNIT 5.4 | NAME __ COURSE __   DATE __ | DRAWING 5.7 |

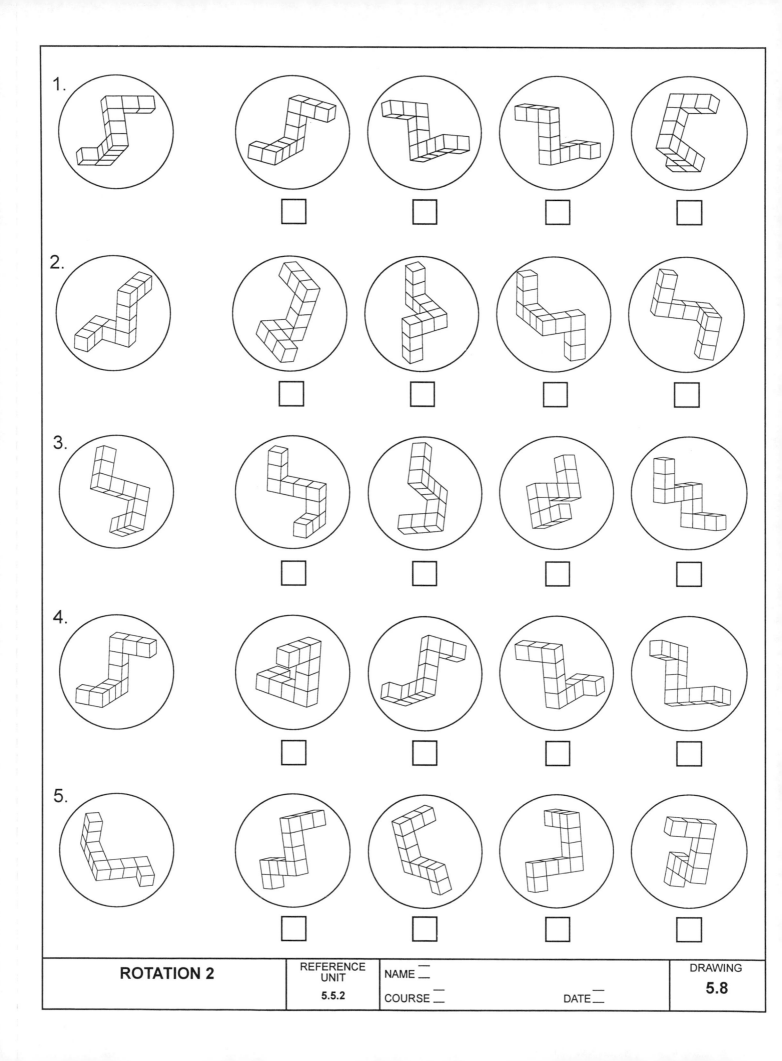

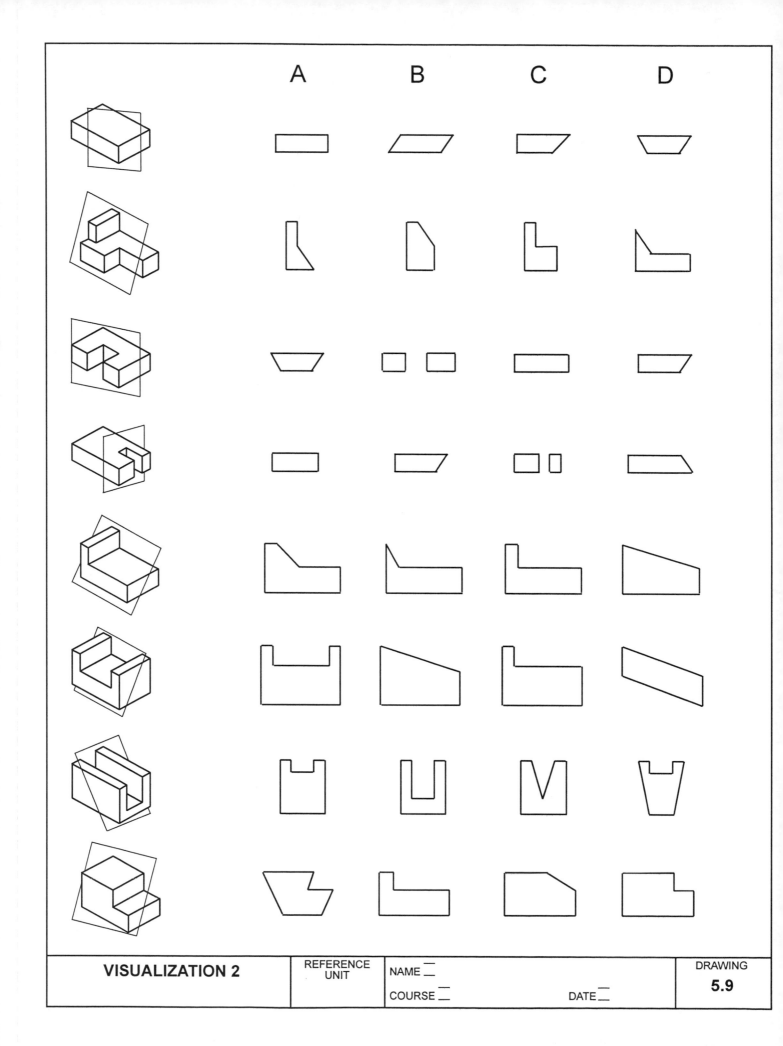

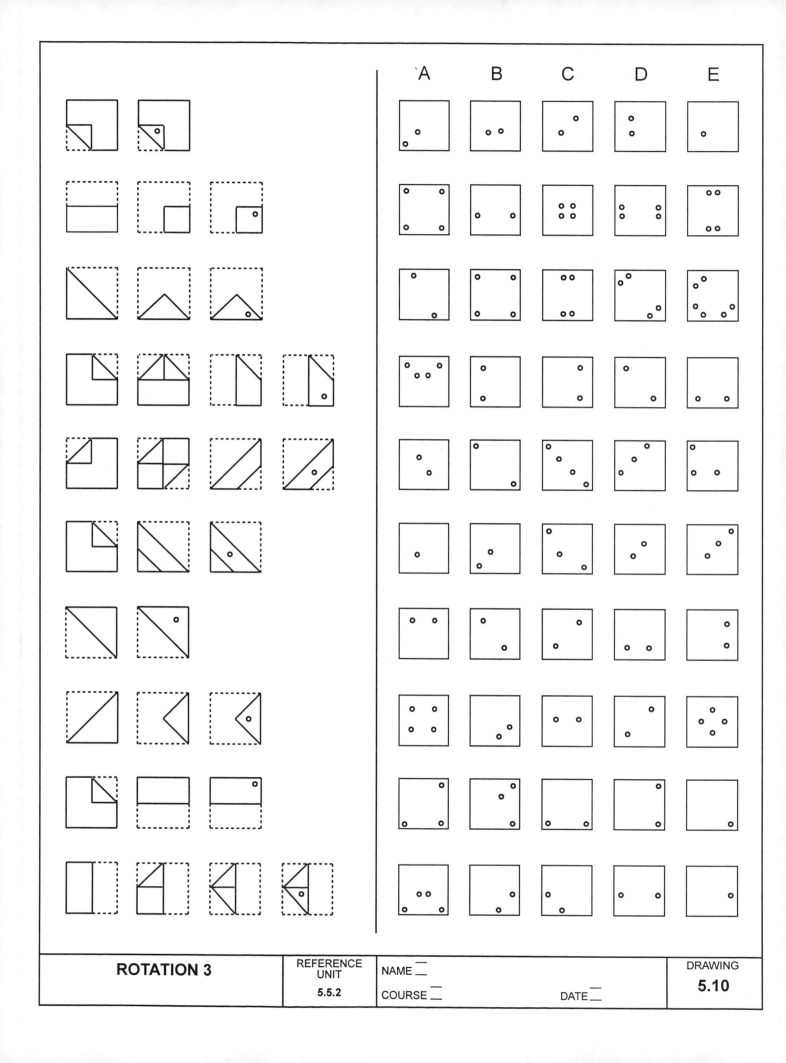

ROTATION 3

REFERENCE UNIT

5.5.2

NAME __

COURSE __

DATE __

DRAWING

5.10

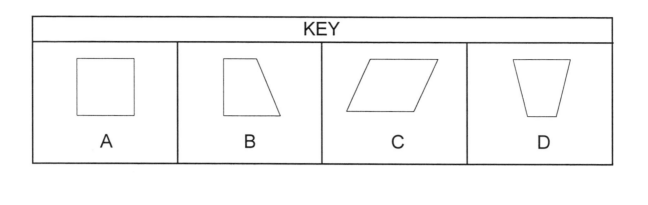

| A | B | C | D |

1. 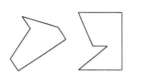 A B C D

2. 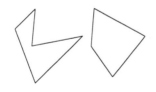 A B C D

3. 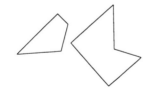 A B C D

4.  A B C D

5. A B C D

6. 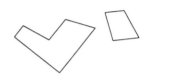 A B C D

7. 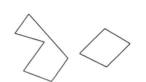 A B C D

8.  A B C D

9. 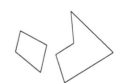 A B C D

10. 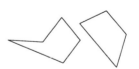 A B C D

| **SPACE RELATIONS** | REFERENCE UNIT | NAME __
 COURSE __ DATE __ | DRAWING
 5.11 |

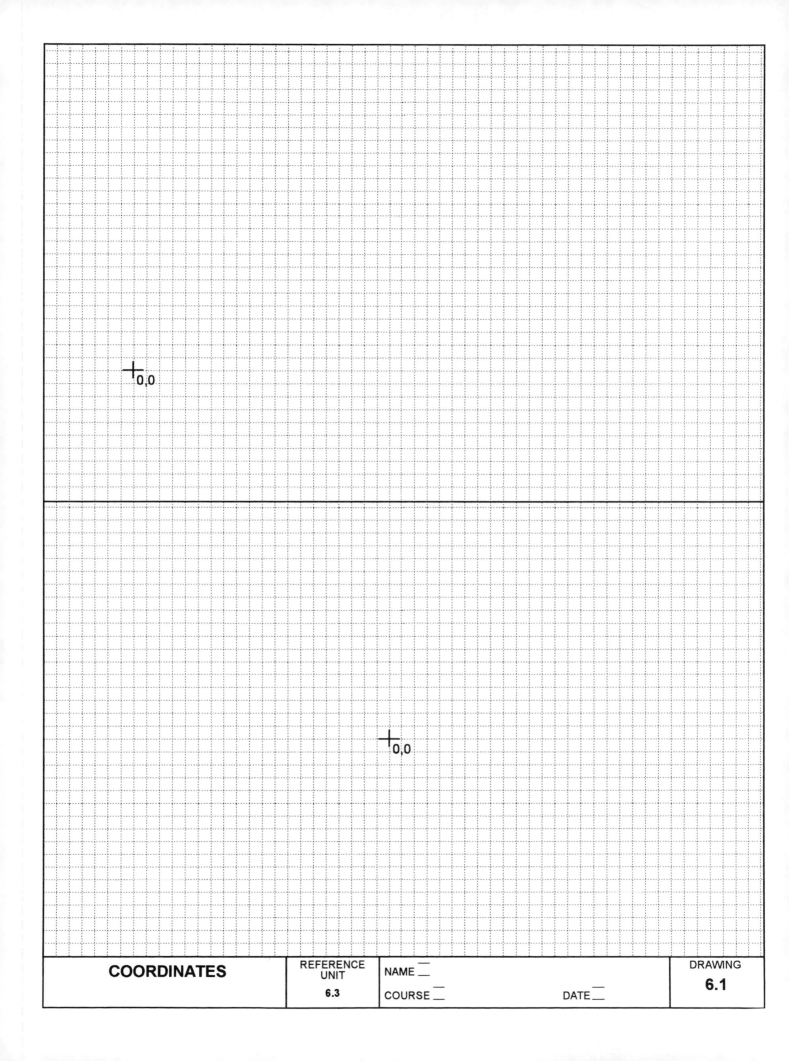

COORDINATES

REFERENCE
UNIT

6.3

NAME __

COURSE __   DATE __

DRAWING

6.1

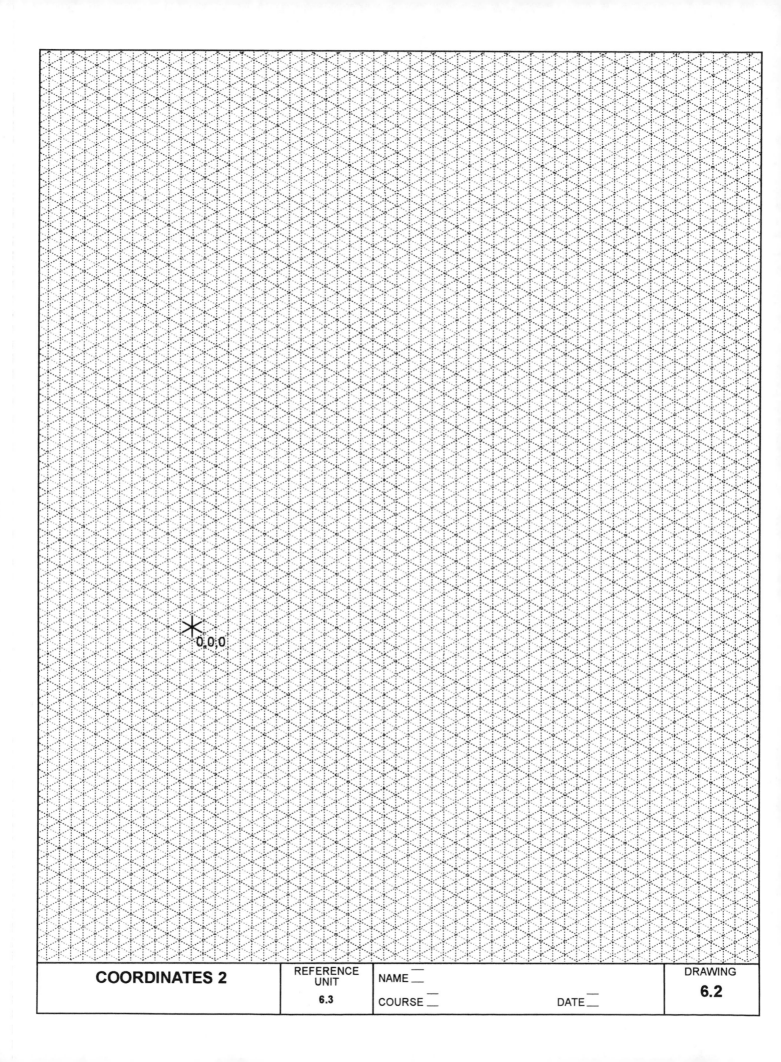

0,0,0

COORDINATES 2

REFERENCE
UNIT

6.3

NAME __

COURSE __   DATE __

DRAWING

6.2

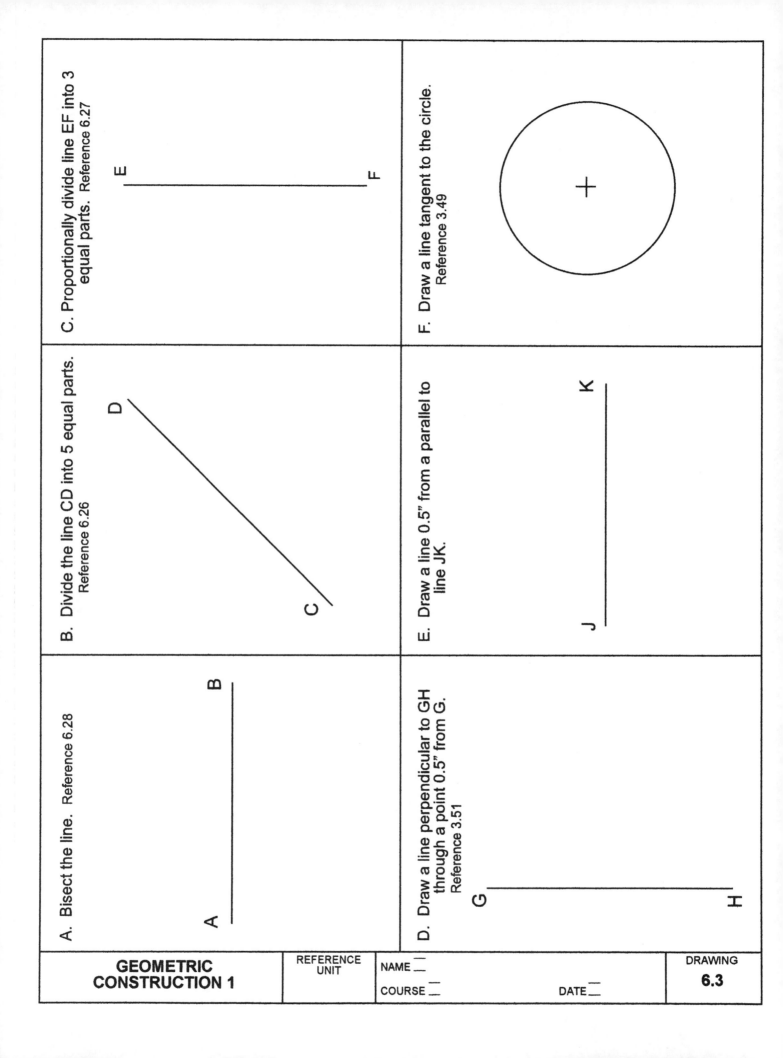

A. Bisect the line. Reference 6.28

A ————— B

B. Divide the line CD into 5 equal parts.
Reference 6.26

C ————— D

C. Proportionally divide line EF into 3
equal parts. Reference 6.27

E ————— F

D. Draw a line perpendicular to GH
through a point 0.5" from G.
Reference 3.51

G ————— H

E. Draw a line 0.5" from a parallel to
line JK.

J ————— K

F. Draw a line tangent to the circle.
Reference 3.49

| GEOMETRIC CONSTRUCTION 1 | REFERENCE UNIT | NAME __ COURSE __ | DATE __ | DRAWING 6.3 |

C. Draw a 1" radius arc tangent to the given lines. Reference 6.37A

F. Construct a 1" radius arc tangent to the two circles.

B. Construct a 1" diameter circle tangent to the given circle. Locate and label the point of tangency. Reference 6.35B

E. Construct a 0.5" radius arc tangent to the circle and the line. Reference 6.38

A. Construct a line tangent to the arc. Locate and label the point of tangency. Reference: 6.35A

D. Draw a 0.5" arc tangent to the given lines. Reference 6.37B

| GEOMETRIC CONSTRUCTION 2 | REFERENCE UNIT | NAME ___ COURSE ___ DATE ___ | DRAWING 6.4 |

C. Rectify the given arc. Reference 6.51

F. Locate two foci by constructing a short perpendicular line 0.5 from each end of the given horizontal line. Construct an ellipse using any technique. Reference 6.68-6.71

B. Draw an ogee curve between the two given lines. Reference 6.48

E. Construct a hyperbola using the equilateral method. Reference 6.54

A. Draw a circle through points A, B, and C. Reference 6.45

+ A

+ B

+ C

D. Construct a parabola using the tangent method. Reference 6.54

A. Construct a spiral of Archimedes.
Reference 6.76

+

B. Construct an involute of the given circle. Reference 6.82

+

C. Bisect the given angle. Reference 6.95

D. Transfer the given angle.
Reference 6.96

E. Construct a square given one of its sides. Reference 6.103

F. Circumscribe a circle around the given triangle.

GEOMETRIC CONSTRUCTION 4

REFERENCE UNIT

NAME __

COURSE __

DATE __

DRAWING

6.6

A. Given the circle, inscribe a square. Reference 6.105

B. Given the circle, circumscribe a square. Reference 6.106

C. Given sides A, B, and C, construct a triangle. Reference 6.108

A _____
B _____
C _____

D. Using the given line construct an equilateral triangle. Reference 6.109

E. Given the circle, inscribe a pentagon. Reference 6.111

F. Given the circle, circumscribe a hexagon. Reference 6.112

GEOMETRIC CONSTRUCTION 5

REFERENCE UNIT

NAME __
COURSE __
DATE __

DRAWING
6.7

A. Given the circle, inscribe a hexagon.
Reference 6.113

B. Given the circle, circumscribe an octagon. Reference 6.114

C. Given the circle, inscribe an octagon.
Reference 6.115

D. Construct a cycloid. Reference 6.78

E. Construct an epicycloid.
Reference 6.79

F. Construct a hypocycloid.
Reference 6.78

| GEOMETRIC CONSTRUCTION 6 | REFERENCE UNIT | NAME __ COURSE __ | DATE __ | DRAWING 6.8 |

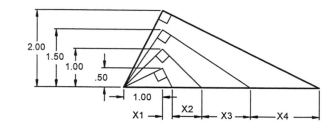

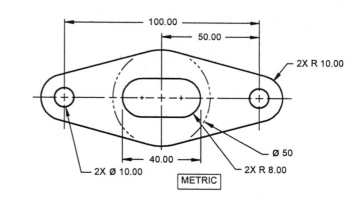

RIDGE GASKET	REFERENCE UNIT	NAME __		DRAWING
		COURSE __	DATE __	6.10

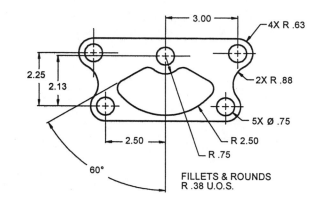

CENTERING PLATE

REFERENCE
UNIT

NAME __

COURSE __

DATE __

DRAWING

6.11

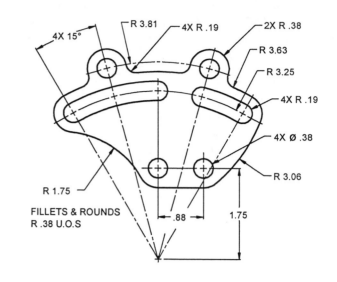

4X 15°   R 3.81   4X R .19   2X R .38

R 3.63

R 3.25

4X R .19

4X Ø .38

R 3.06

R 1.75

FILLETS & ROUNDS
R .38 U.O.S

.88   1.75

| ARCHED FOLLOWER | REFERENCE UNIT | NAME __ | | DRAWING |
| | | COURSE __   DATE __ | | 6.12 |

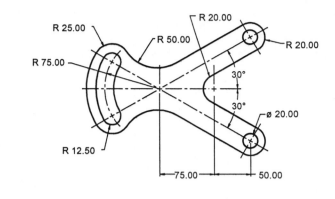

| WING PLATE | REFERENCE UNIT | NAME __ | | DRAWING |
| | | COURSE __ | DATE __ | 6.13 |

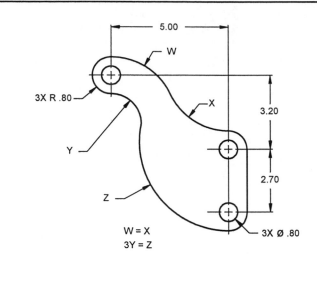

5.00

W

3X R .80

X

3.20

Y

2.70

Z

W = X
3Y = Z

3X Ø .80

| TRANSITION | REFERENCE UNIT | NAME __
COURSE __ | DATE __ | DRAWING
6.14 |

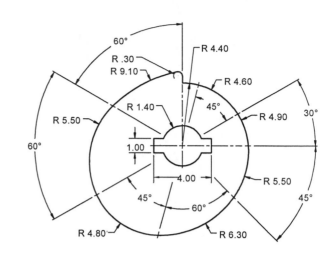

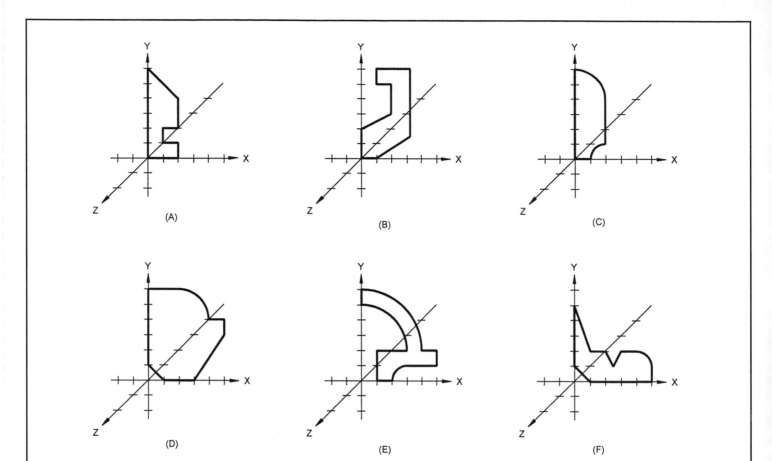

CIRCULAR SWEEP 1

REFERENCE UNIT

7.6.4

NAME __

COURSE __

DATE __

DRAWING

7.1

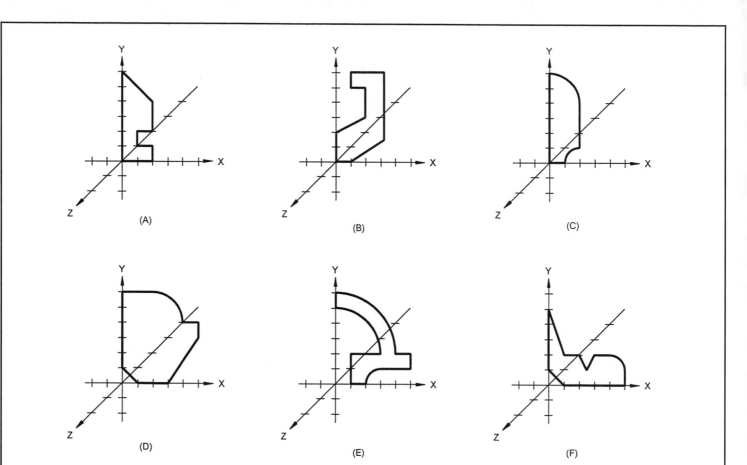

CIRCULAR SWEEP 2

REFERENCE
UNIT

7.6.4

NAME —

COURSE —

DATE —

DRAWING

7.2

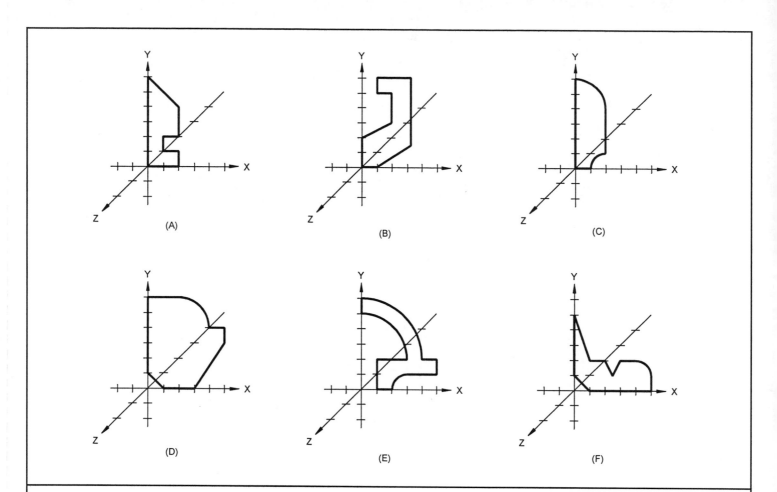

(A)   (B)   (C)

(D)   (E)   (F)

| LINEAR SWEEP | REFERENCE UNIT
7.6.4 | NAME —
COURSE — | DATE — | DRAWING
7.3 |

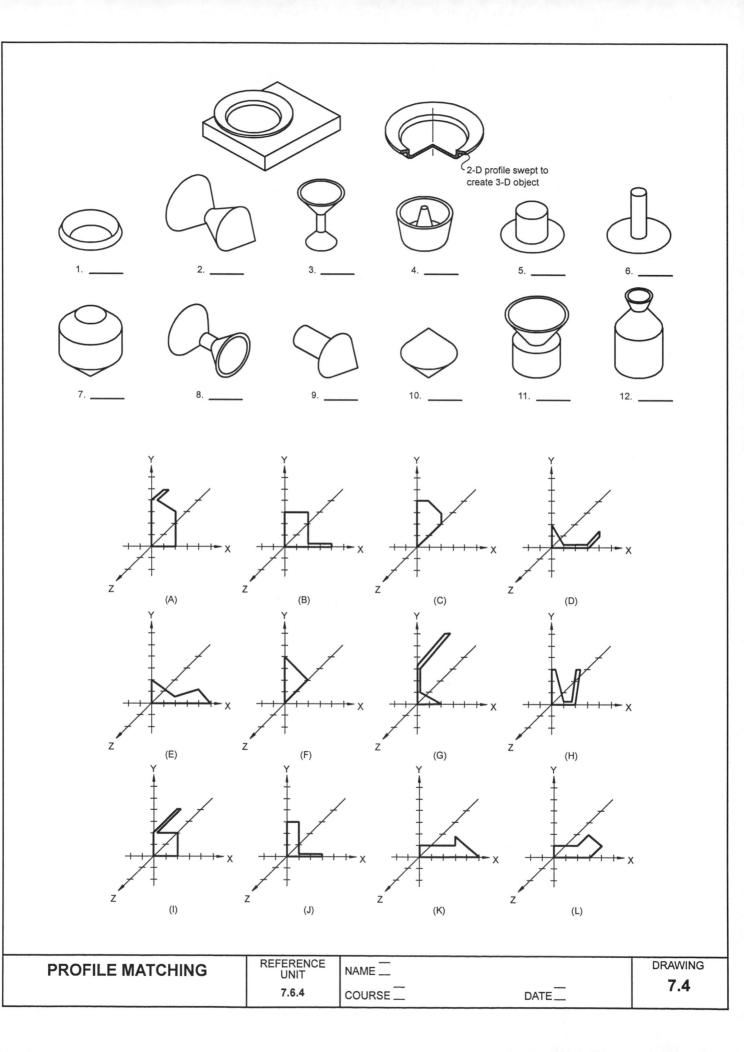

2-D profile swept to create 3-D object

1. ___
2. ___
3. ___
4. ___
5. ___
6. ___

7. ___
8. ___
9. ___
10. ___
11. ___
12. ___

(A) (B) (C) (D)

(E) (F) (G) (H)

(I) (J) (K) (L)

A. (A U B) U C

B. (A U B) -C

C. (A - B) -C

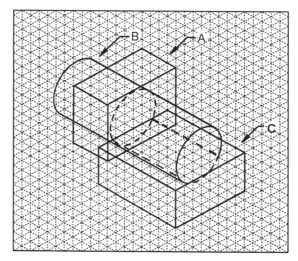

BOOLEAN OPERATIONS 1	REFERENCE UNIT	NAME __		DRAWING
	7.4.2	COURSE __	DATE __	**7.5**

A. (A - B) -C

B. (A U B)U C

C. B- (A U C)

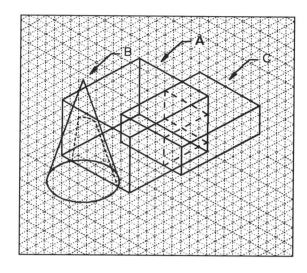

A. (C - A) -B

B. (A U C) -B

C. (A int C) -B

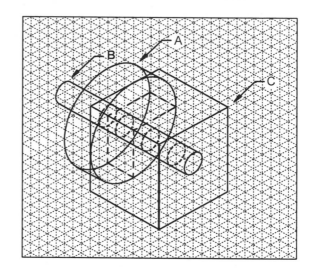

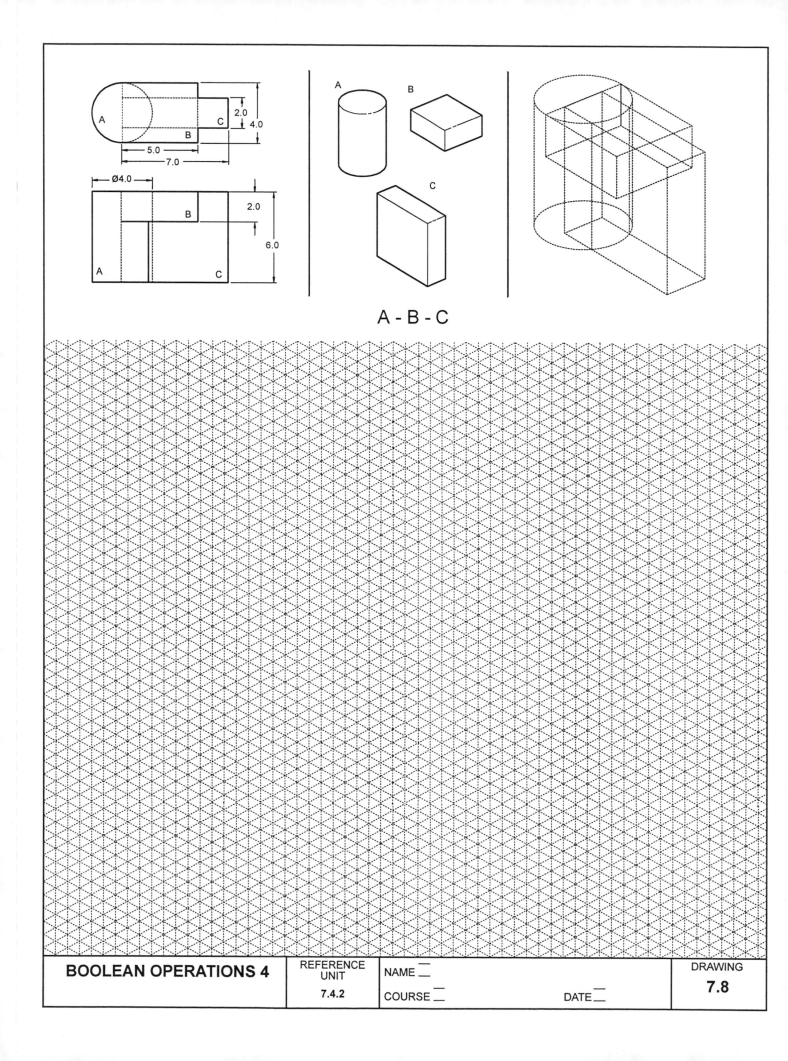

A - B - C

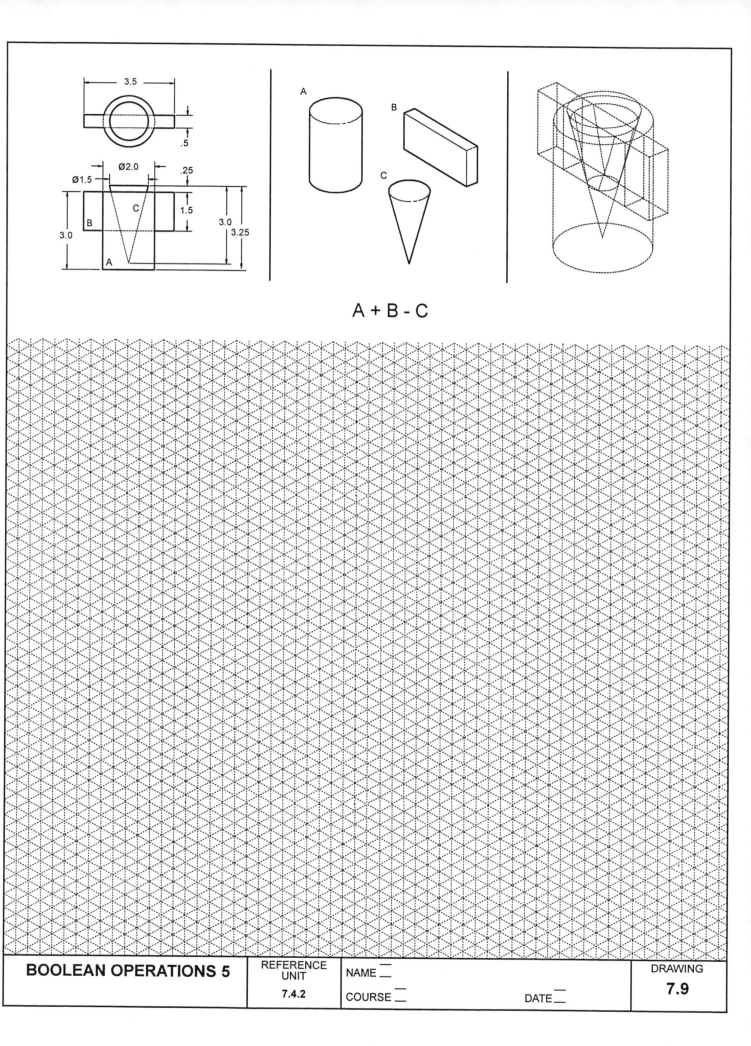

3.5

.5

Ø2.0   .25

Ø1.5

C

B   1.5   3.0   3.0   3.25

A

A

B

C

A + B - C

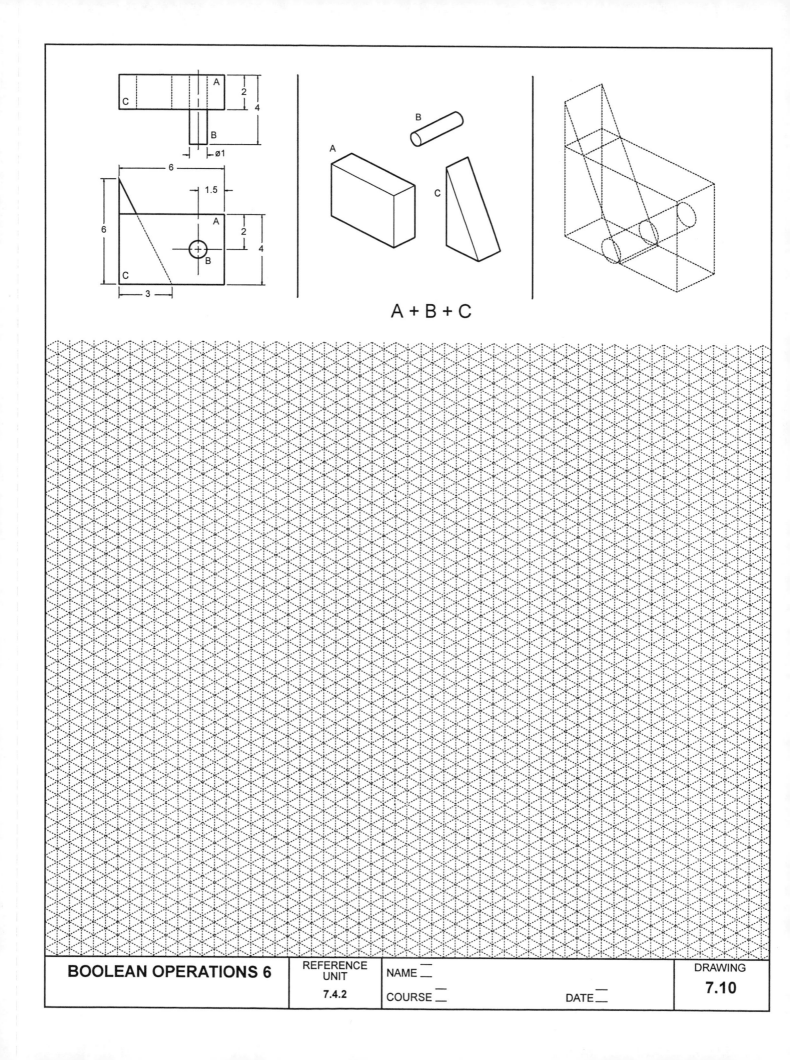

A + B + C

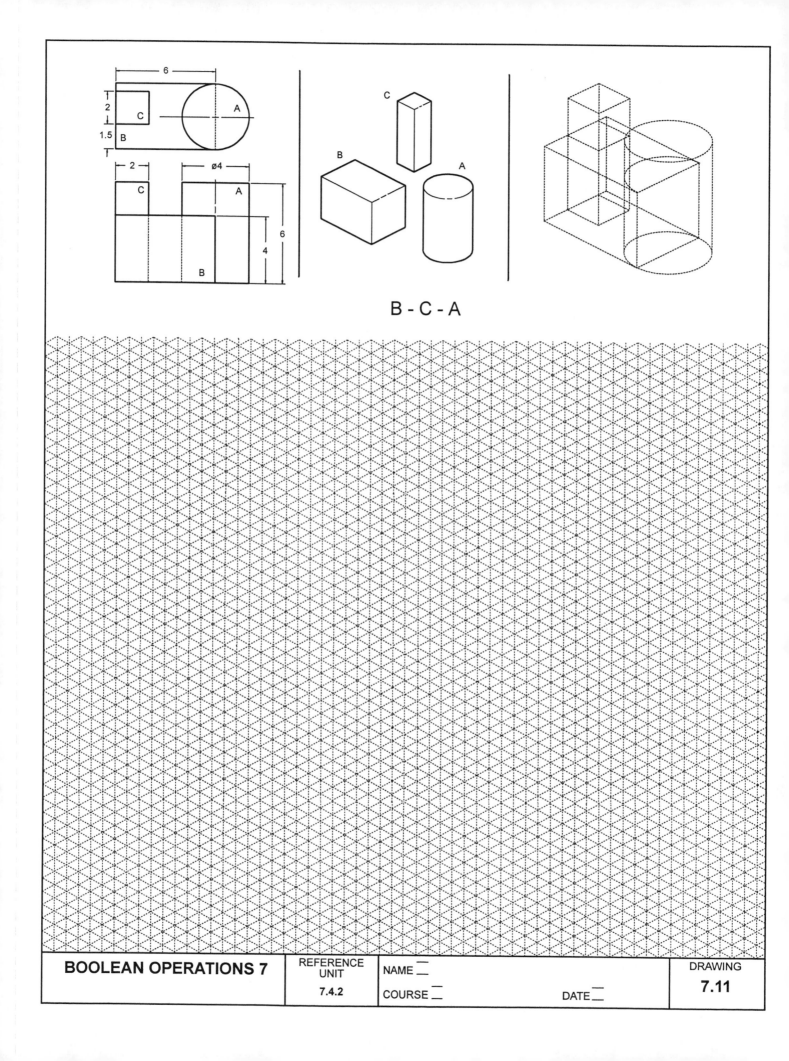

B - C - A

BOOLEAN OPERATIONS 7

REFERENCE UNIT

7.4.2

NAME __

COURSE __

DATE __

DRAWING

7.11

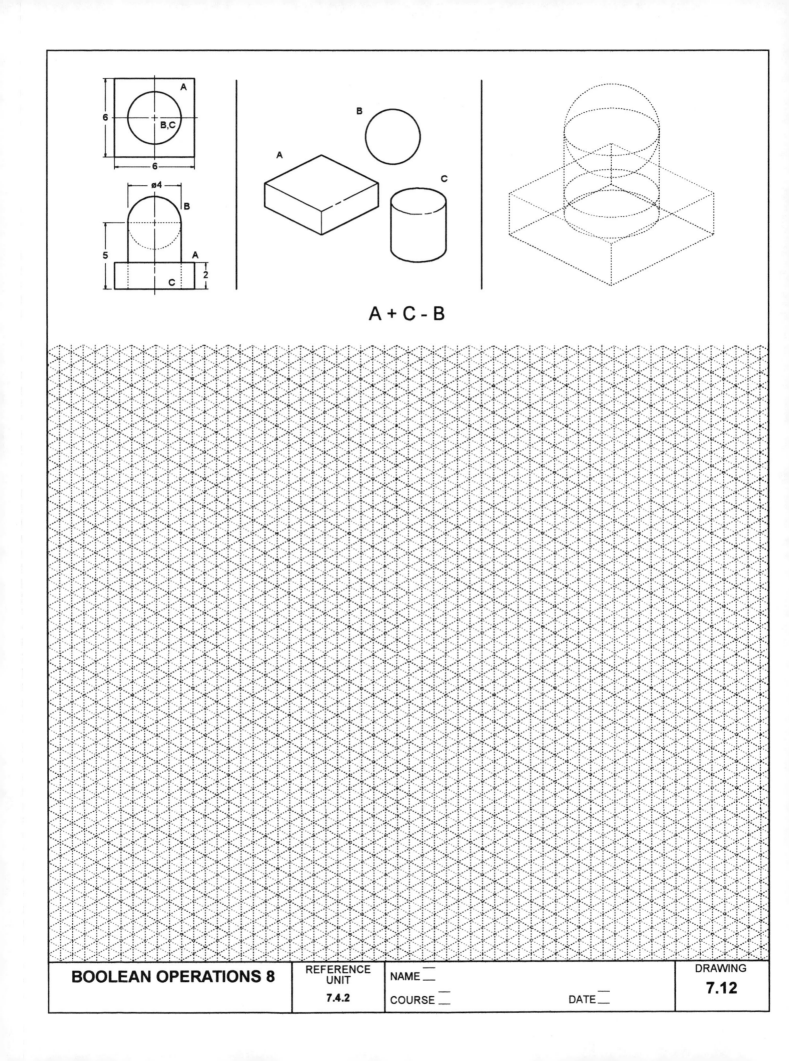

A + C - B

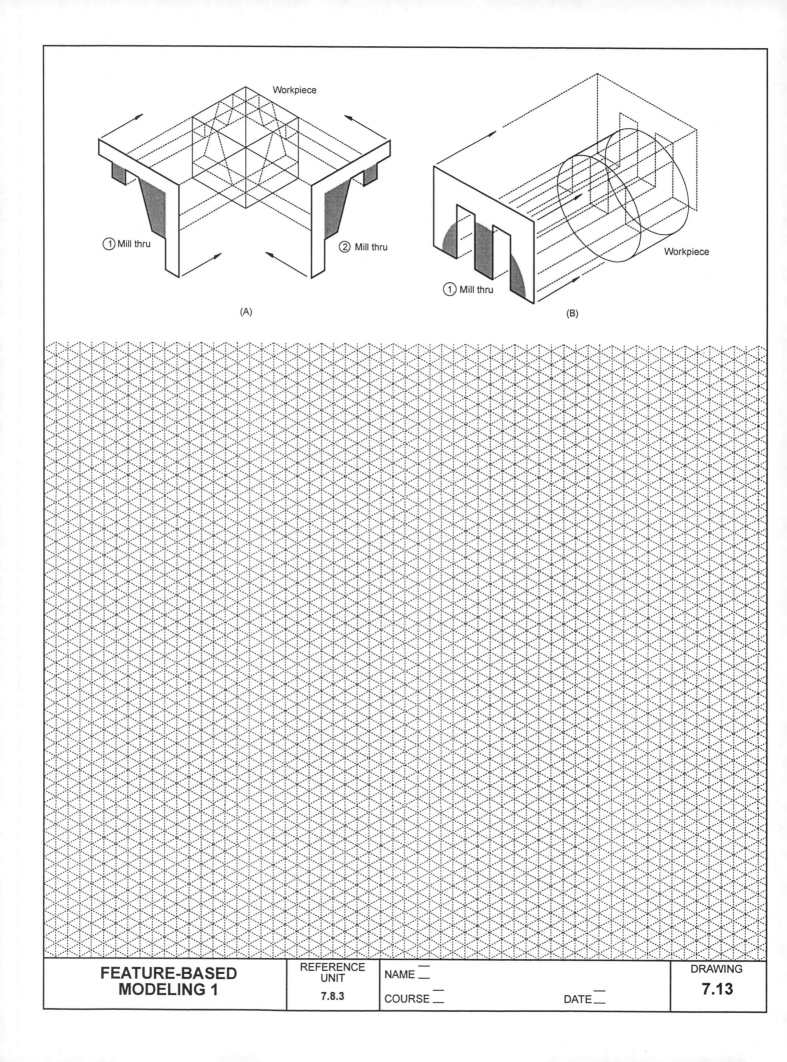

Workpiece

① Mill thru

② Mill thru

(A)

① Mill thru

Workpiece

(B)

| FEATURE-BASED MODELING 1 | REFERENCE UNIT 7.8.3 | NAME __ COURSE __ DATE __ | DRAWING 7.13 |

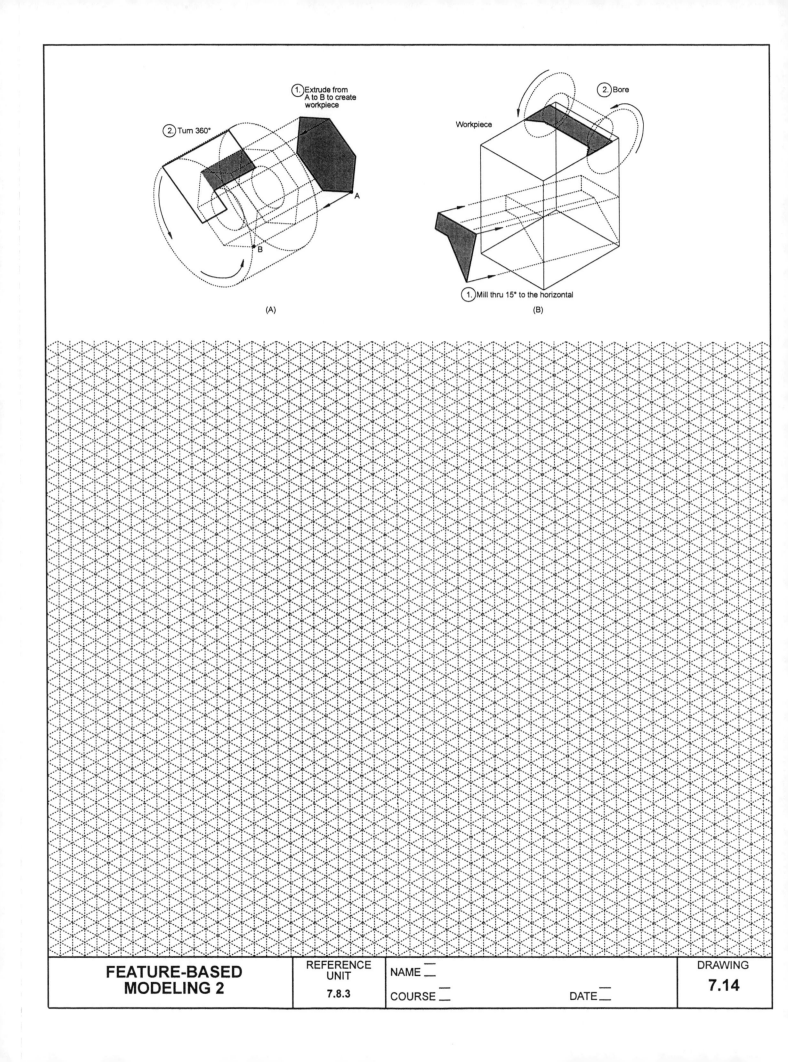

2. Turn 360°

1. Extrude from
A to B to create
workpiece

A

B

(A)

Workpiece

2. Bore

1. Mill thru 15° to the horizontal

(B)

| FEATURE-BASED MODELING 2 | REFERENCE UNIT 7.8.3 | NAME __ COURSE __ DATE __ | DRAWING 7.14 |

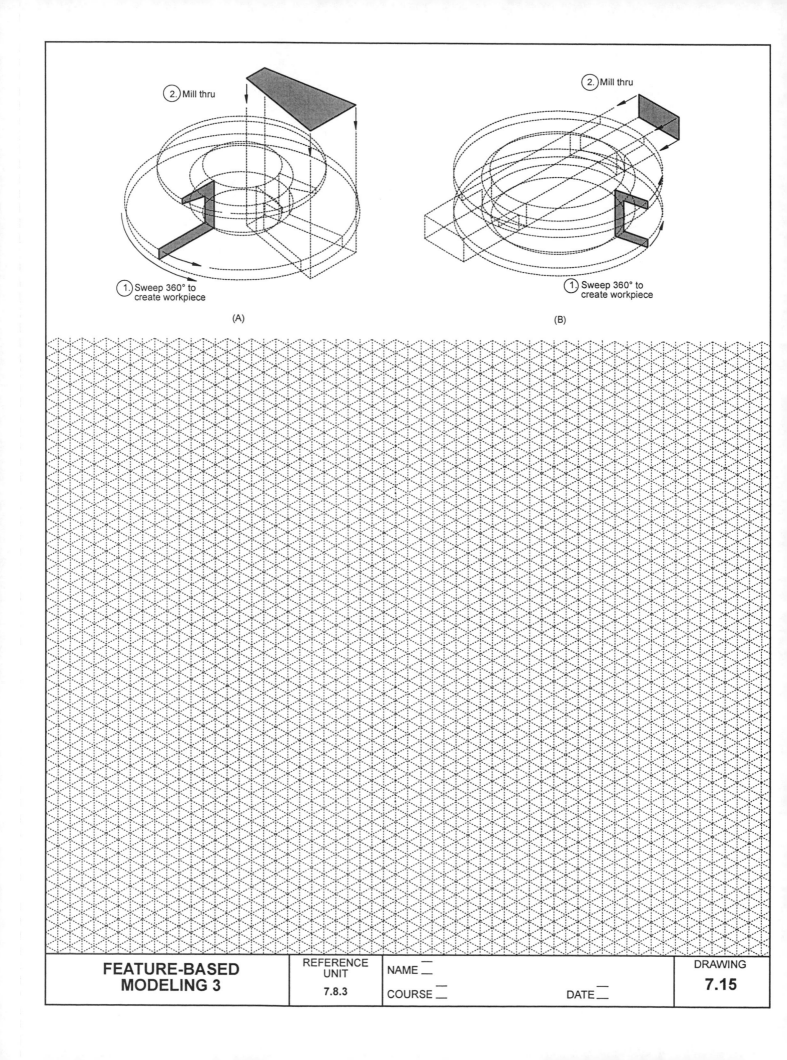

2. Mill thru

1. Sweep 360° to create workpiece

(A)

2. Mill thru

1. Sweep 360° to create workpiece

(B)

| FEATURE-BASED MODELING 3 | REFERENCE UNIT 7.8.3 | NAME __ COURSE __ DATE __ | DRAWING 7.15 |

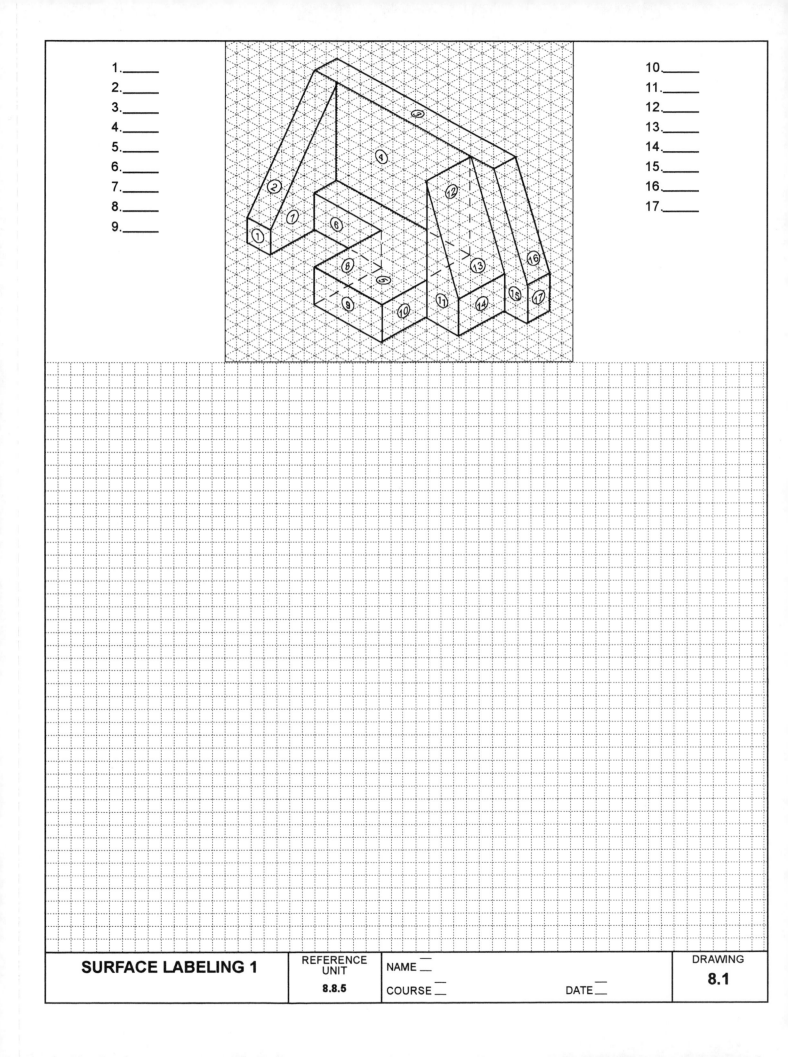

1.____
2.____
3.____
4.____
5.____
6.____
7.____
8.____
9.____

10.____
11.____
12.____
13.____
14.____
15.____
16.____
17.____

SURFACE LABELING 1

REFERENCE UNIT
8.8.5

NAME __

COURSE __

DATE __

DRAWING
8.1

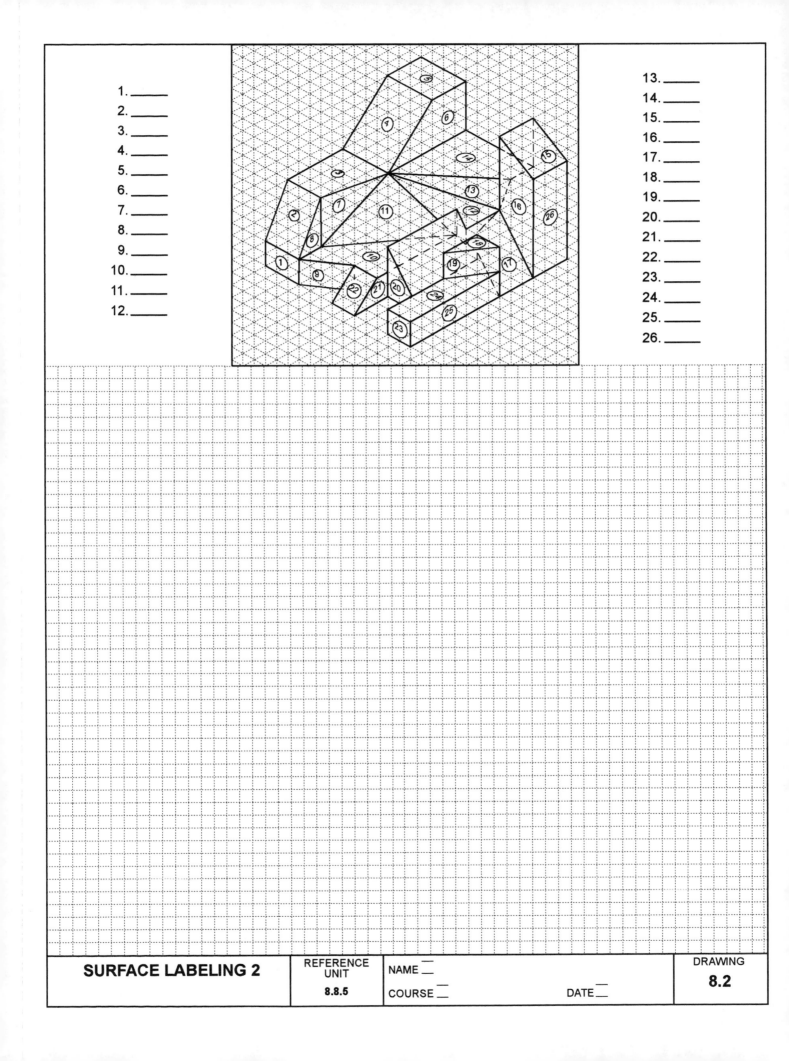

SURFACE LABELING 2

REFERENCE UNIT

8.8.5

NAME ___

COURSE ___

DATE ___

DRAWING

8.2

1. ___
2. ___
3. ___
4. ___
5. ___
6. ___
7. ___
8. ___
9. ___
10. ___
11. ___
12. ___

13. ___
14. ___
15. ___
16. ___
17. ___
18. ___
19. ___
20. ___
21. ___
22. ___
23. ___
24. ___
25. ___
26. ___

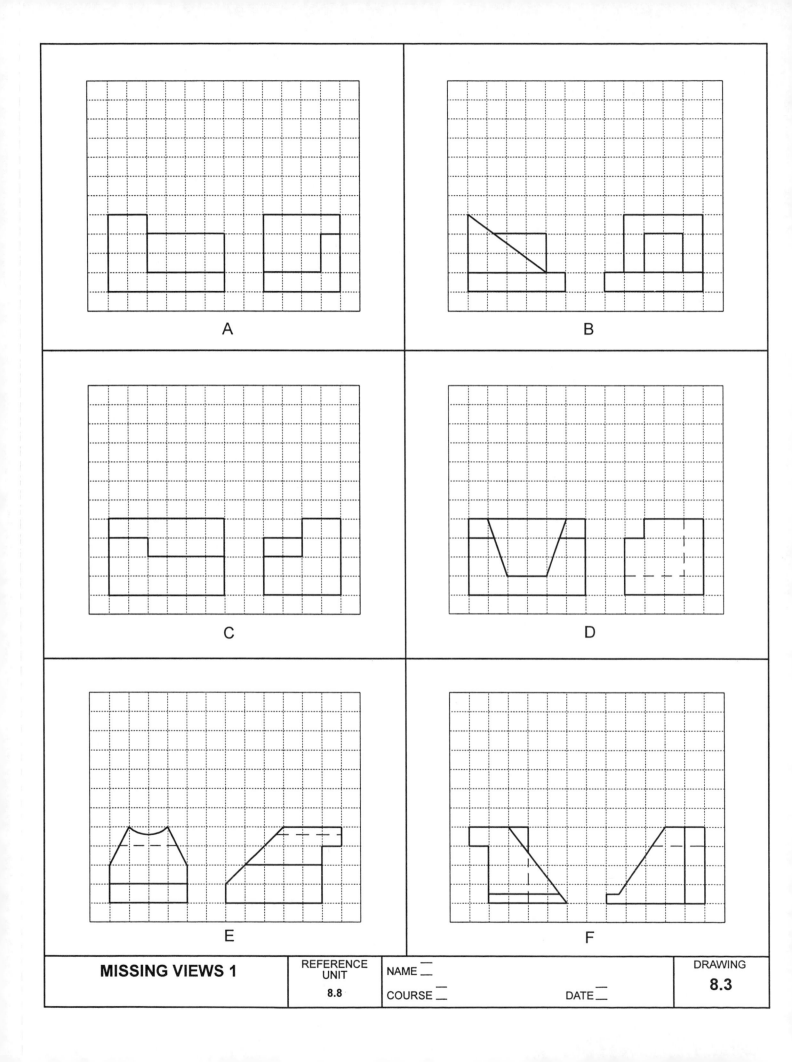

A

B

C

D

E

F

| **MISSING VIEWS 1** | REFERENCE UNIT 8.8 | NAME __ COURSE __ | DATE __ | DRAWING 8.3 |

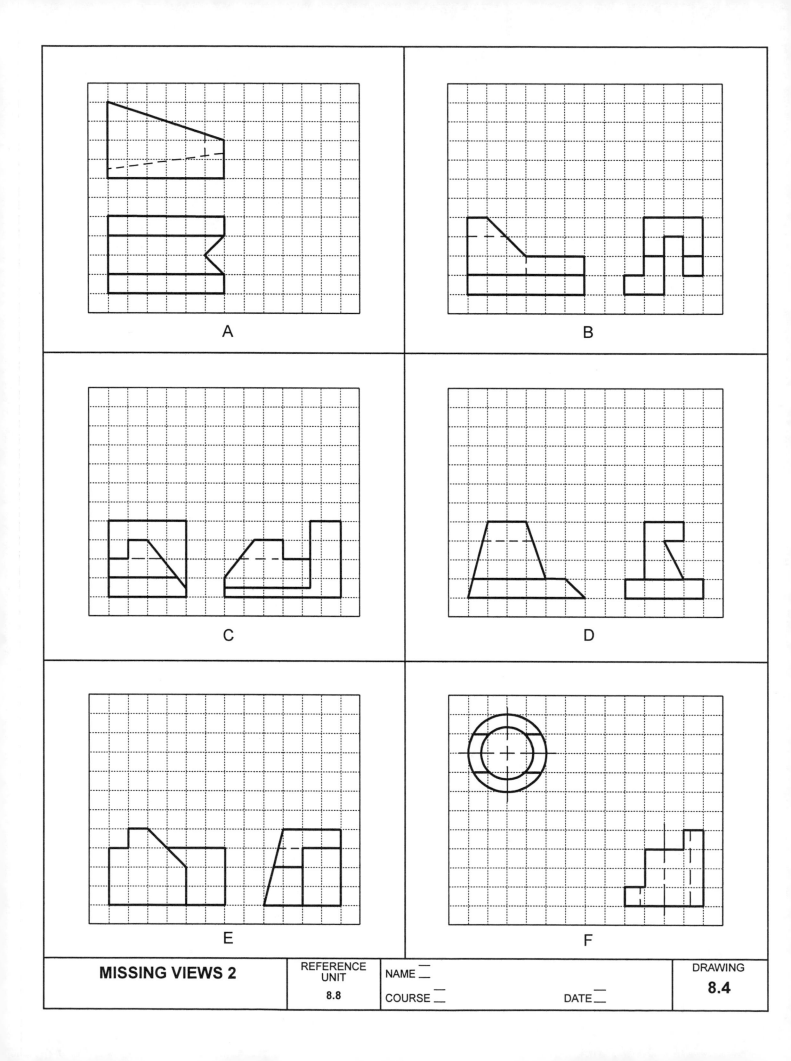

A

B

C

D

E

F

MISSING VIEWS 2

REFERENCE UNIT

8.8

NAME __

COURSE __

DATE __

DRAWING

8.4

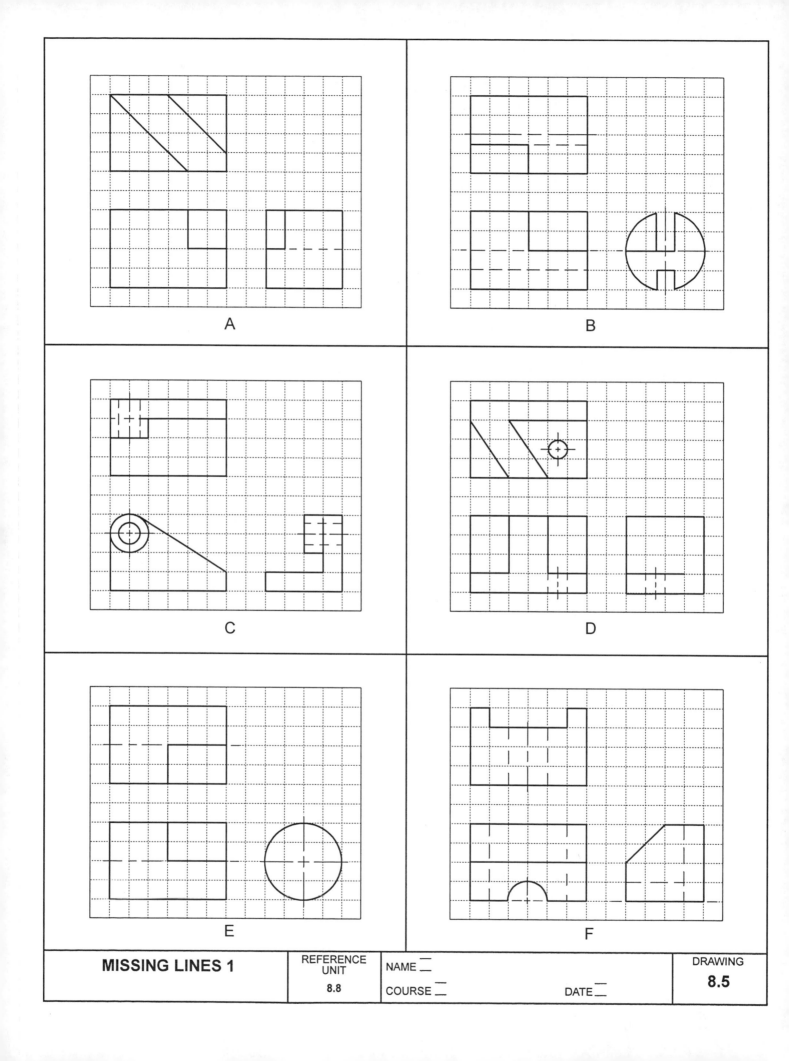

A

B

C

D

E

F

MISSING LINES 1

REFERENCE UNIT

8.8

NAME __

COURSE __

DATE __

DRAWING

8.5

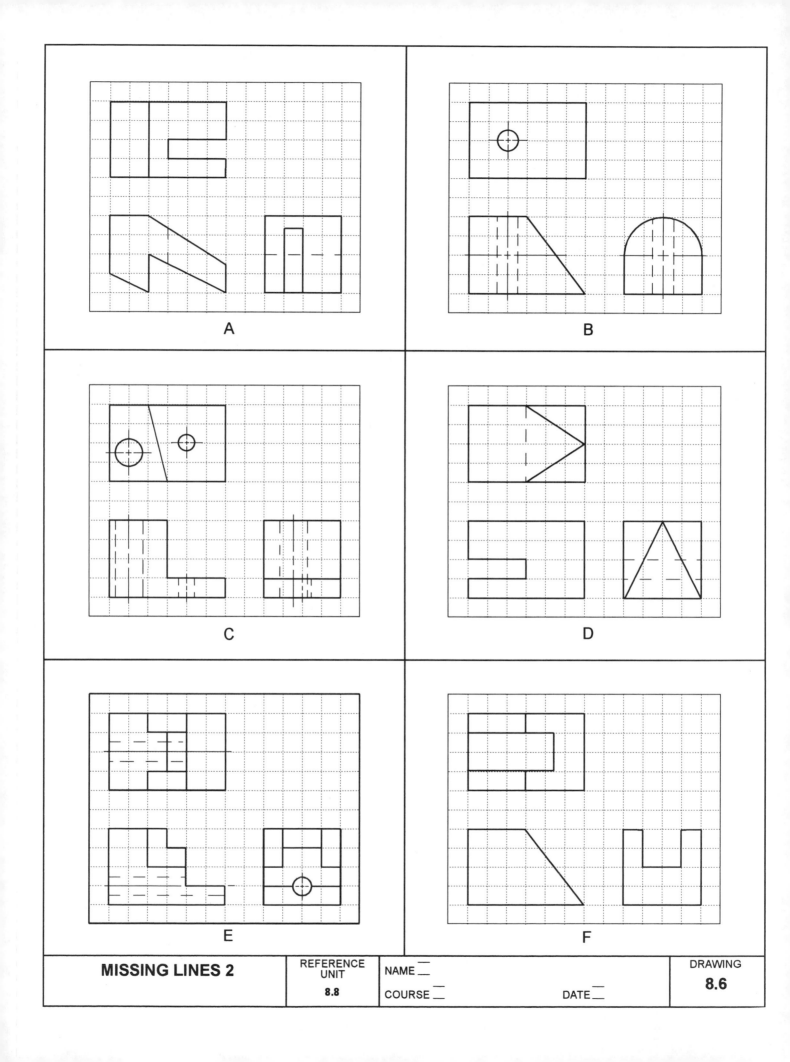

A

B

C

D

E

F

MISSING LINES 2

REFERENCE UNIT

8.8

NAME __

COURSE __

DATE __

DRAWING

8.6

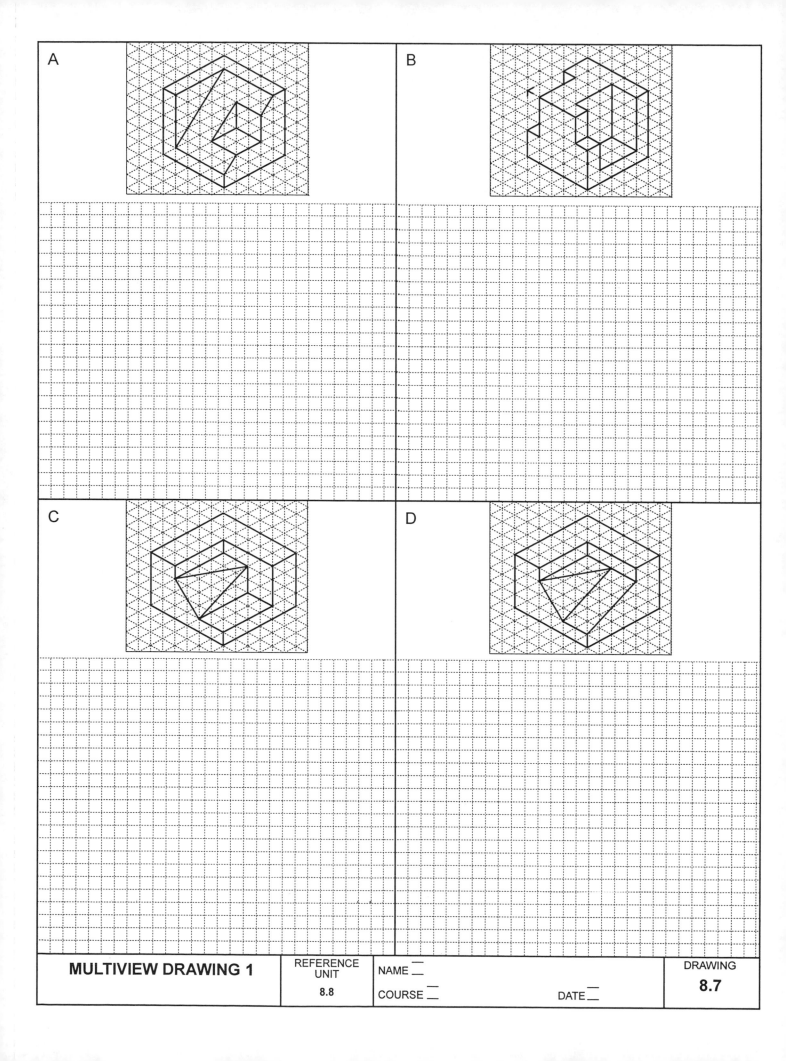

A

B

C

D

MULTIVIEW DRAWING 1

REFERENCE UNIT

8.8

NAME __

COURSE __

DATE __

DRAWING

8.7

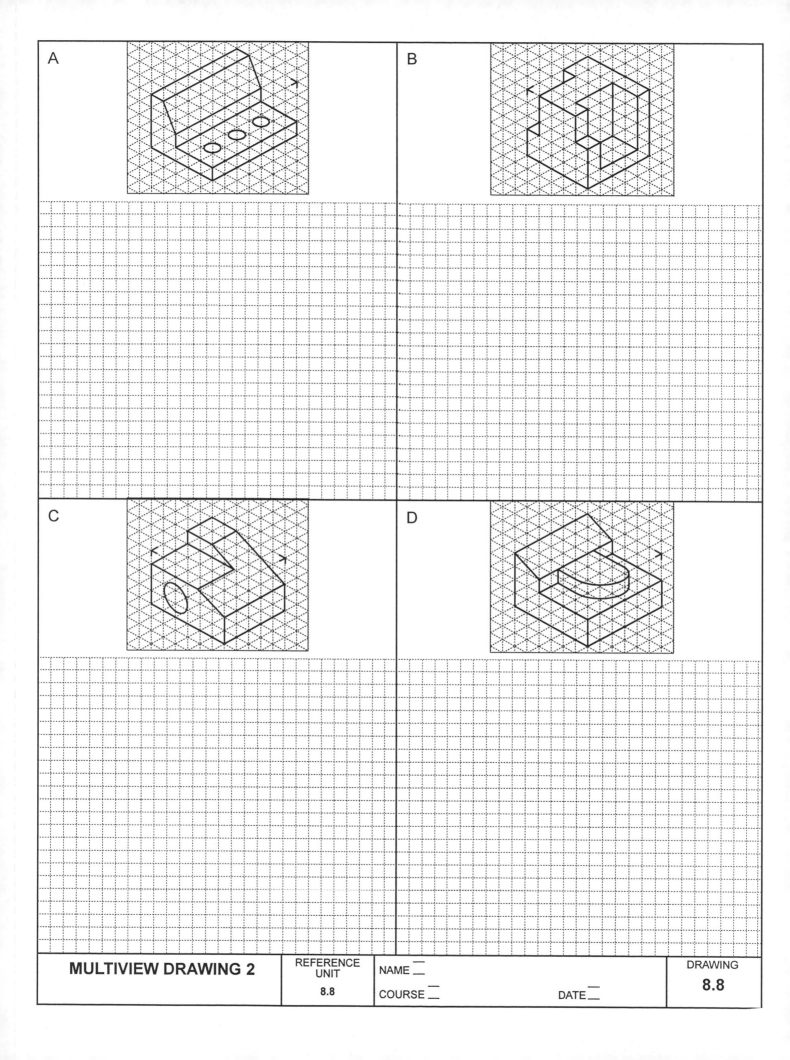

A

B

C

D

MULTIVIEW DRAWING 2

REFERENCE
UNIT

8.8

NAME __

COURSE __

DATE __

DRAWING

8.8

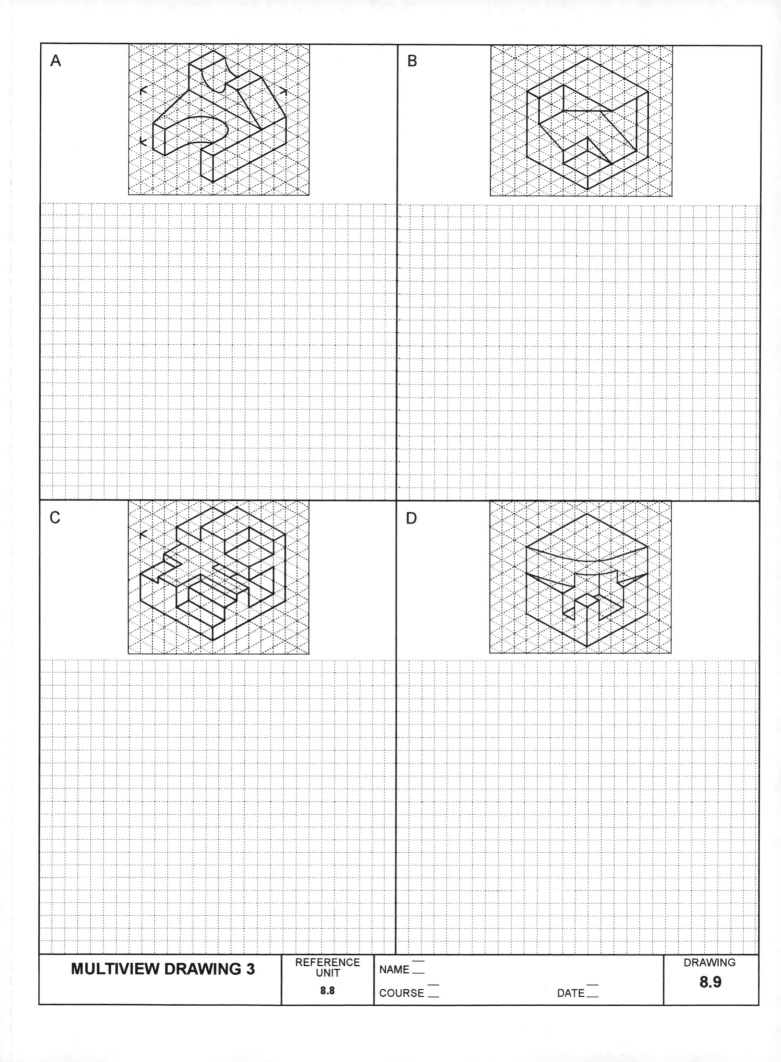

A

B

C

D

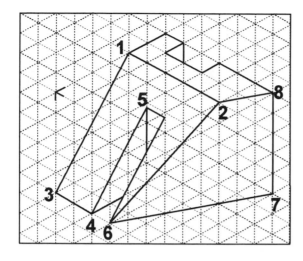

1-2 _____

1-3 _____

3-4 _____

4-5 _____

2-6 _____

6-7 _____

7-8 _____

8-2 _____

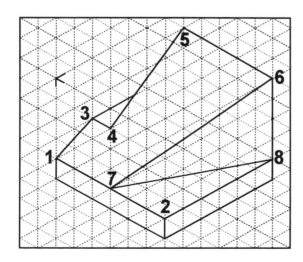

1-2 _____

1-3 _____

3-4 _____

4-5 _____

5-6 _____

6-7 _____

7-8 _____

8-6 _____

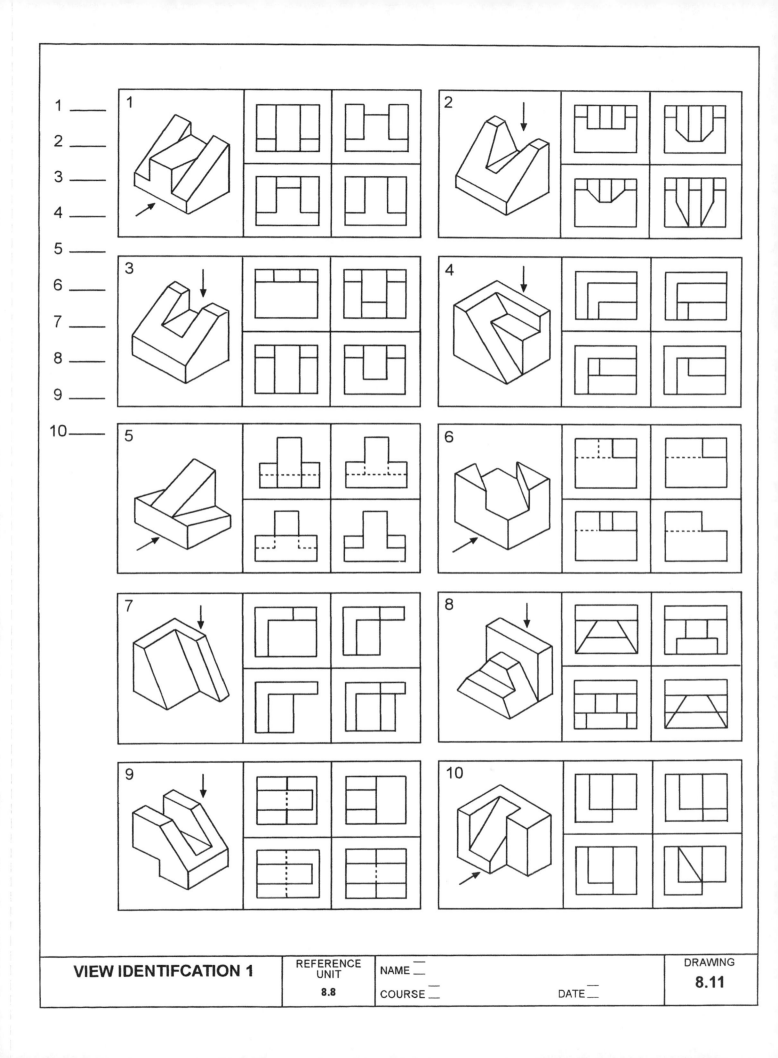

VIEW IDENTIFCATION 1

REFERENCE UNIT

8.8

NAME __

COURSE __

DATE __

DRAWING

8.11

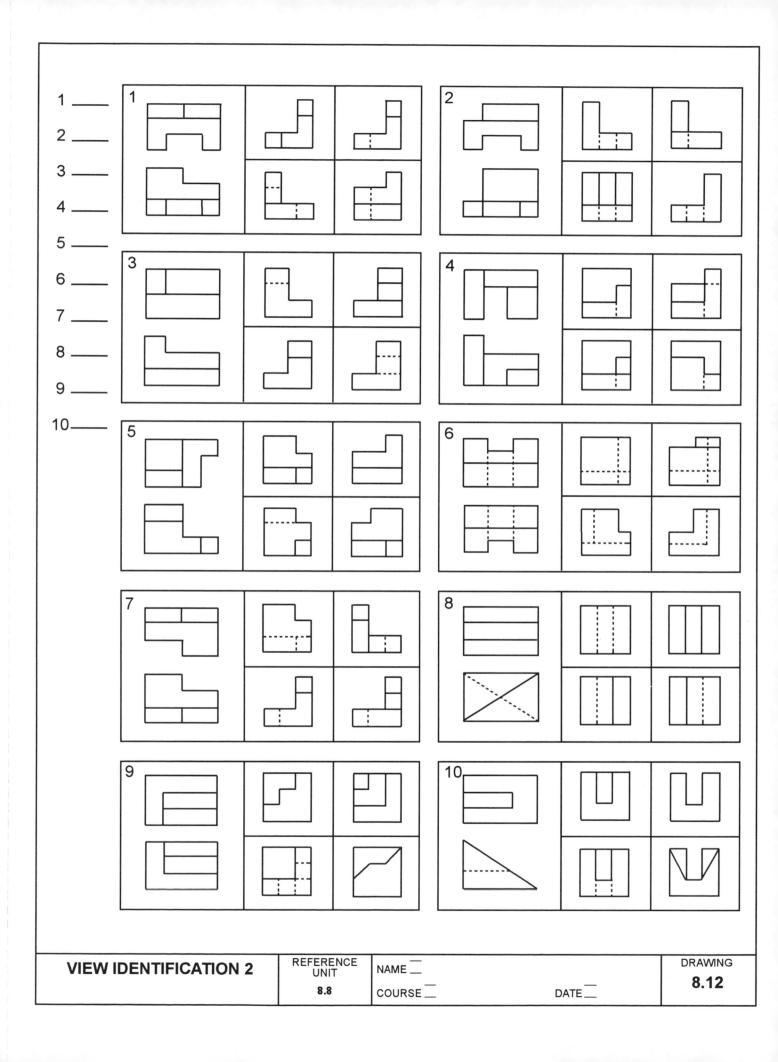

1 ——
2 ——
3 ——
4 ——
5 ——
6 ——
7 ——
8 ——
9 ——
10——

VIEW IDENTIFICATION 2

REFERENCE UNIT
8.8

NAME __
COURSE __
DATE __

DRAWING
8.12

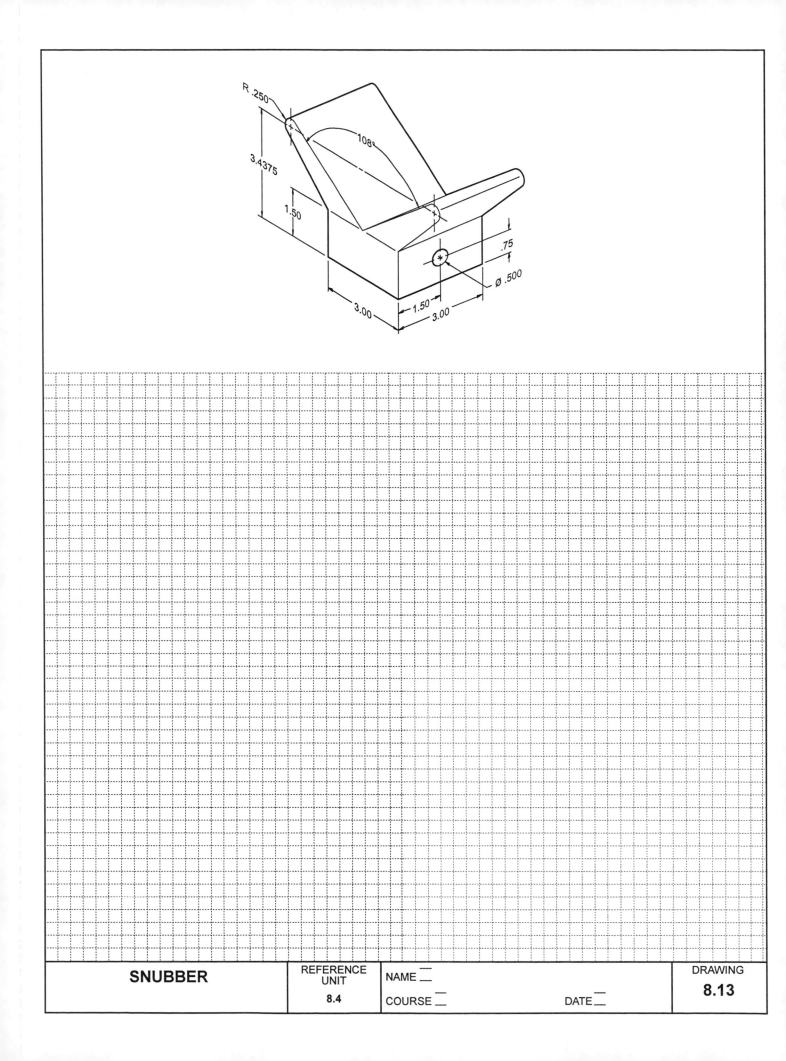

R .250

108°

3.4375

1.50

3.00

1.50

3.00

.75

Ø .500

| SNUBBER | REFERENCE UNIT 8.4 | NAME __ COURSE __ DATE __ | DRAWING 8.13 |

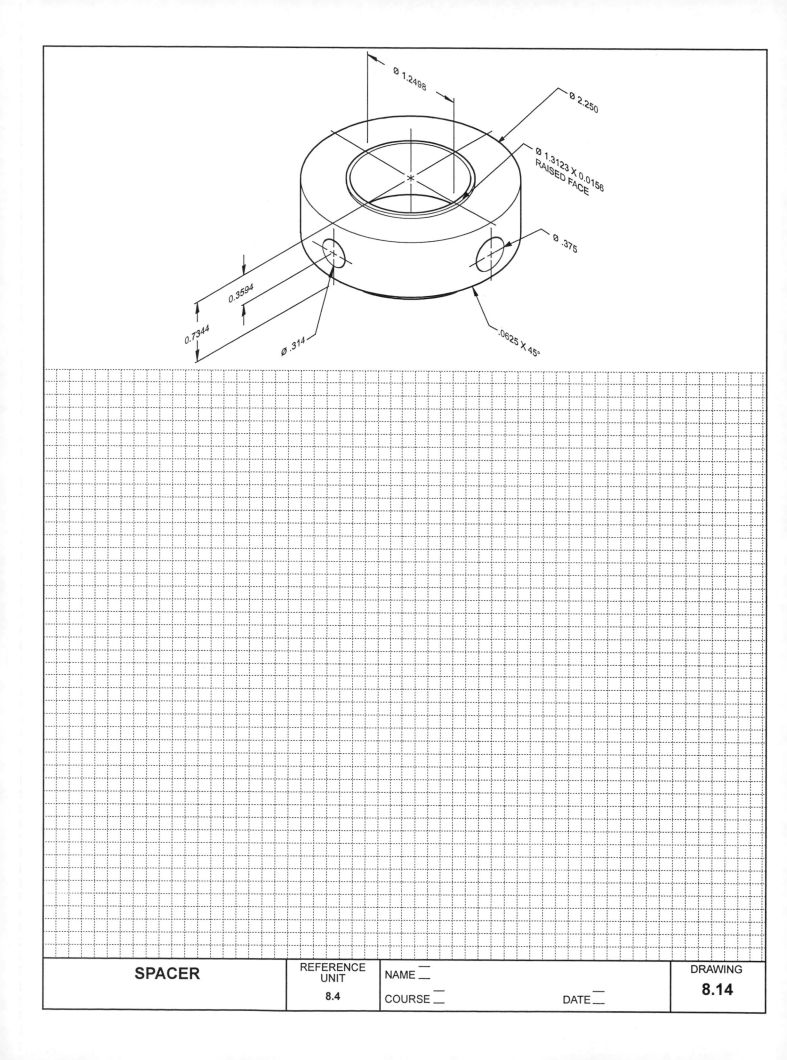

Ø 1.2498

Ø 2.250

Ø 1.3123 X 0.0156
RAISED FACE

Ø .375

0.3594

0.7344

Ø .314

.0625 X 45°

| SPACER | REFERENCE UNIT 8.4 | NAME ⎯ COURSE ⎯ DATE ⎯ | DRAWING 8.14 |

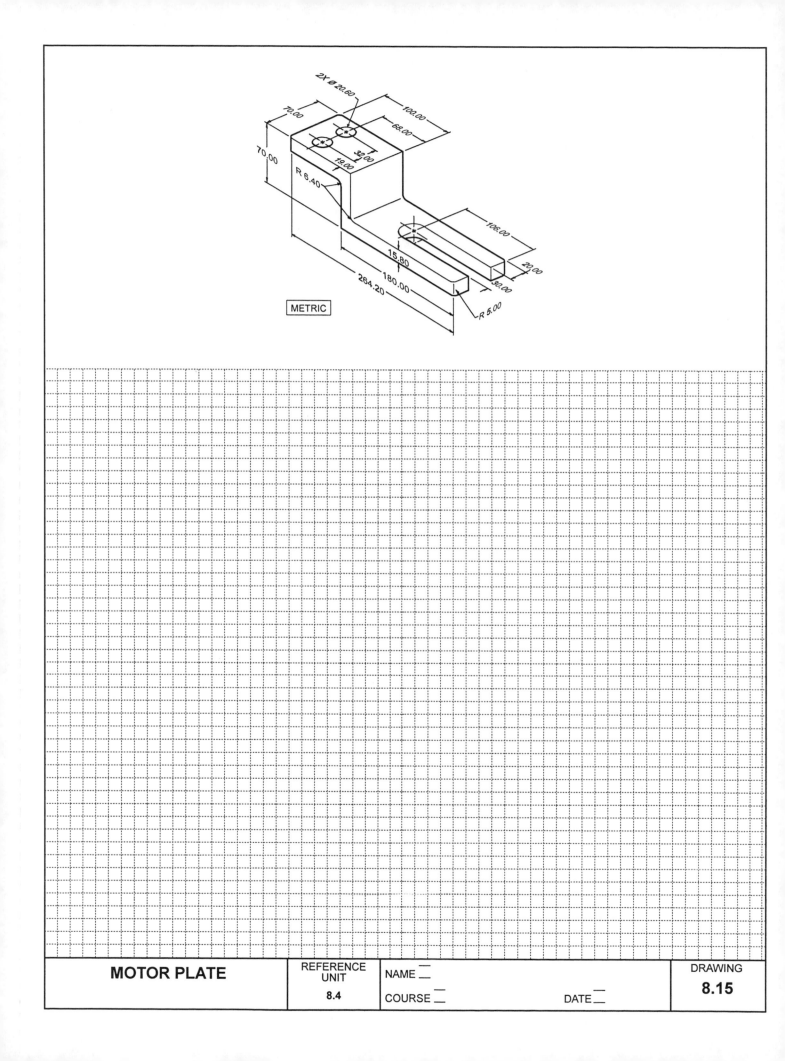

2X Ø 20.60

70.00

100.00

68.00

70.00

32.00

19.00

R 6.40

106.00

15.80

20.00

30.00

264.20

180.00

R 5.00

METRIC

| MOTOR PLATE | REFERENCE UNIT | NAME __ | | DRAWING |
| | 8.4 | COURSE __ | DATE __ | 8.15 |

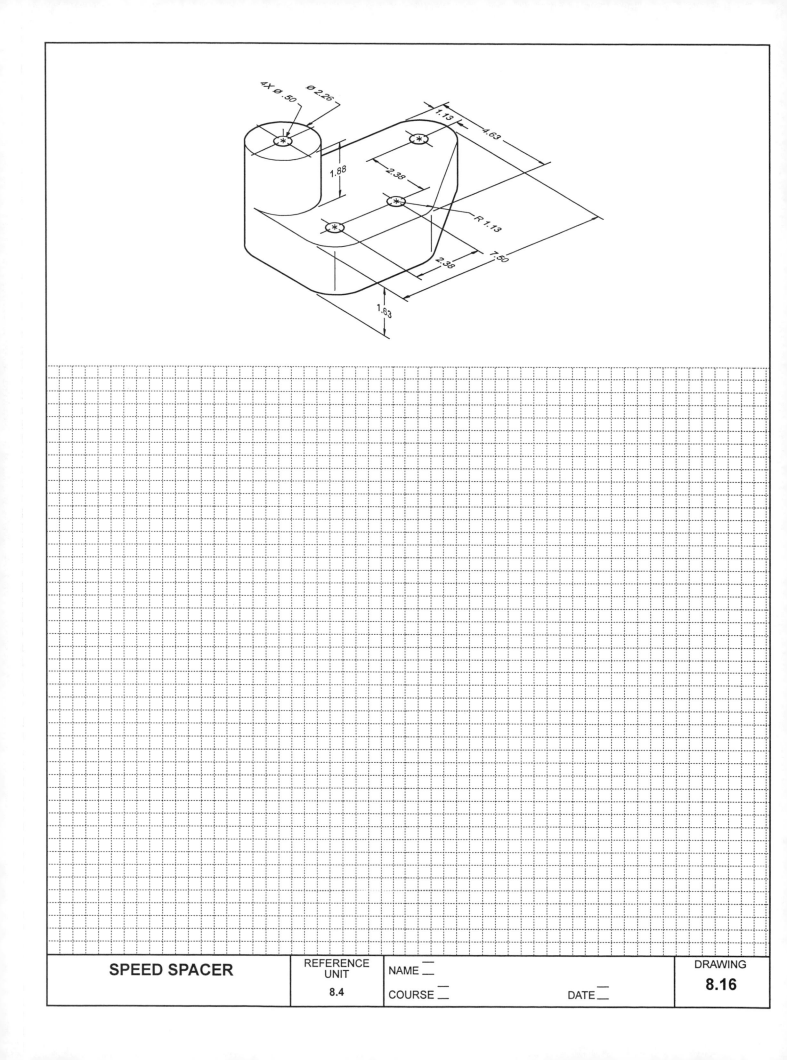

4X Ø.50

Ø 2.26

1.13

4.63

1.88

2.38

R 1.13

2.38

7.50

1.63

| SPEED SPACER | REFERENCE UNIT 8.4 | NAME __ COURSE __ DATE __ | DRAWING 8.16 |

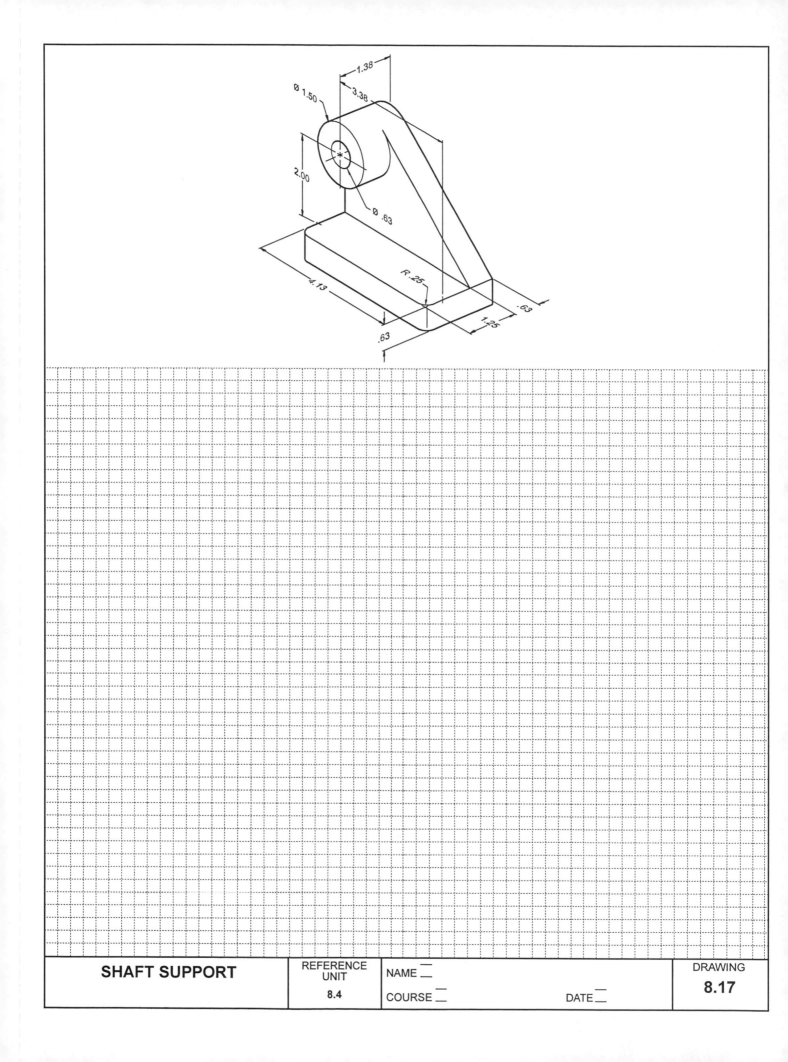

Ø 1.50
1.38
3.38
2.00
Ø .63
R .25
4.13
.63
1.25
.63

SHAFT SUPPORT

REFERENCE UNIT

8.4

NAME __

COURSE __

DATE __

DRAWING

8.17

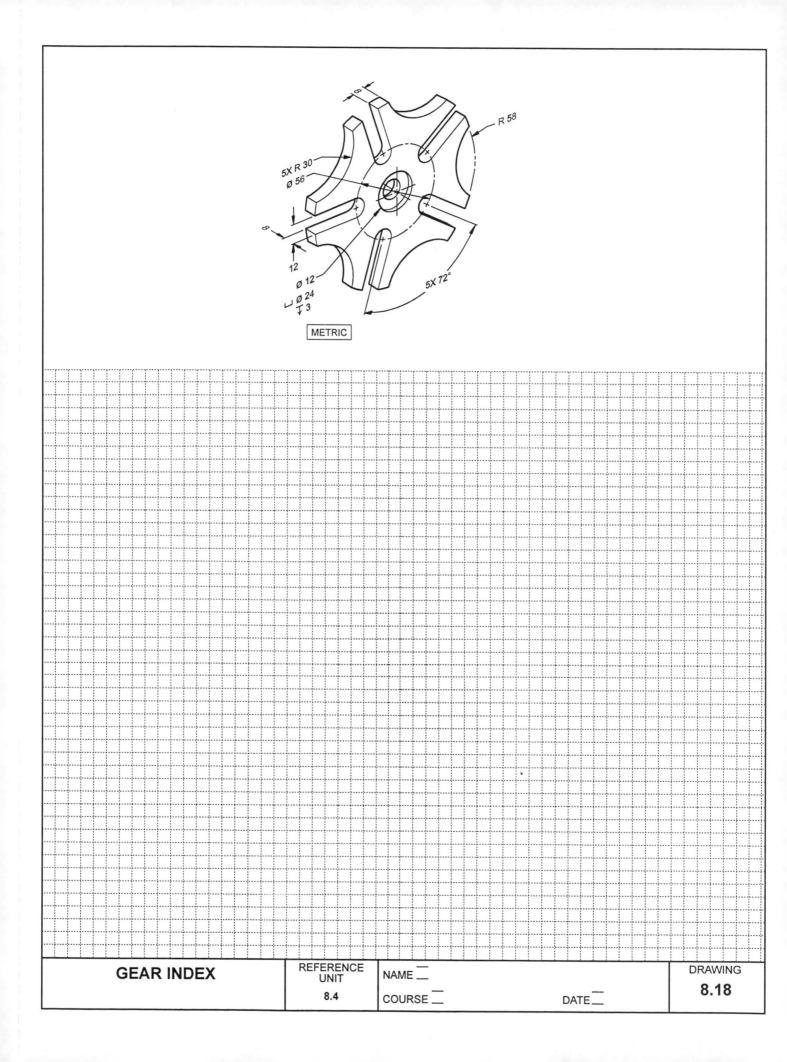

5X R 30
Ø 56
R 58
8
8
12
Ø 12
Ø 24
3
5X 72°
METRIC

| GEAR INDEX | REFERENCE UNIT | NAME __ | | DRAWING |
| | 8.4 | COURSE __ | DATE __ | 8.18 |

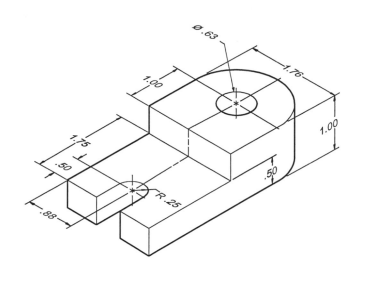

STOP BASE

REFERENCE UNIT

8.4

NAME __

COURSE __

DATE __

DRAWING

8.19

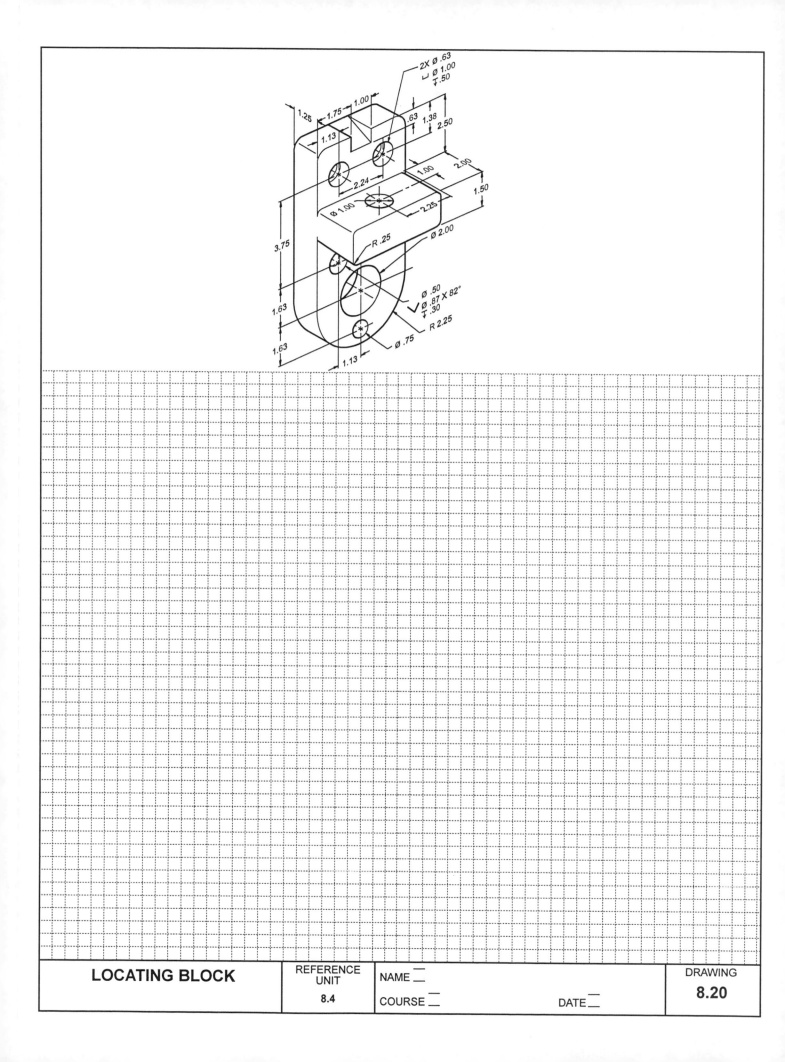

2X Ø .63
⌴ Ø 1.00
▼ .50

1.25
1.75
1.00
1.13
.63
1.38
2.50
2.24
1.00
2.00
Ø 1.00
1.50
2.25
3.75
R .25
Ø 2.00
Ø .50
Ø .87 X 82°
▼ .30
1.63
R 2.25
1.63
Ø .75
1.13

LOCATING BLOCK

REFERENCE UNIT

8.4

NAME __

COURSE __

DATE __

DRAWING

8.20

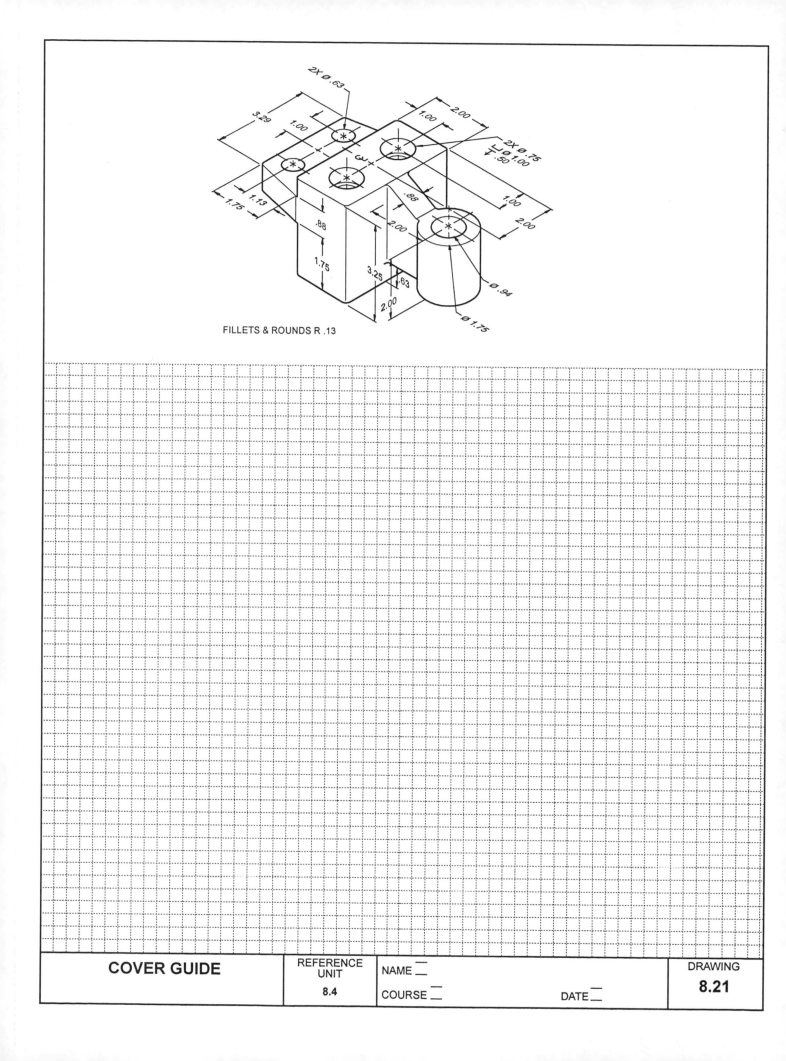

2X ⌀.63

3.29

1.00

2.00

1.00

2X ⌀.75
⊔ ⌀1.00
▽ .50

1.13

1.75

.88

.88

1.00

2.00

2.00

1.75

3.25

2.00

.63

.94

⌀ 1.75

FILLETS & ROUNDS R .13

COVER GUIDE

REFERENCE UNIT

8.4

NAME —

COURSE —

DATE —

DRAWING

8.21

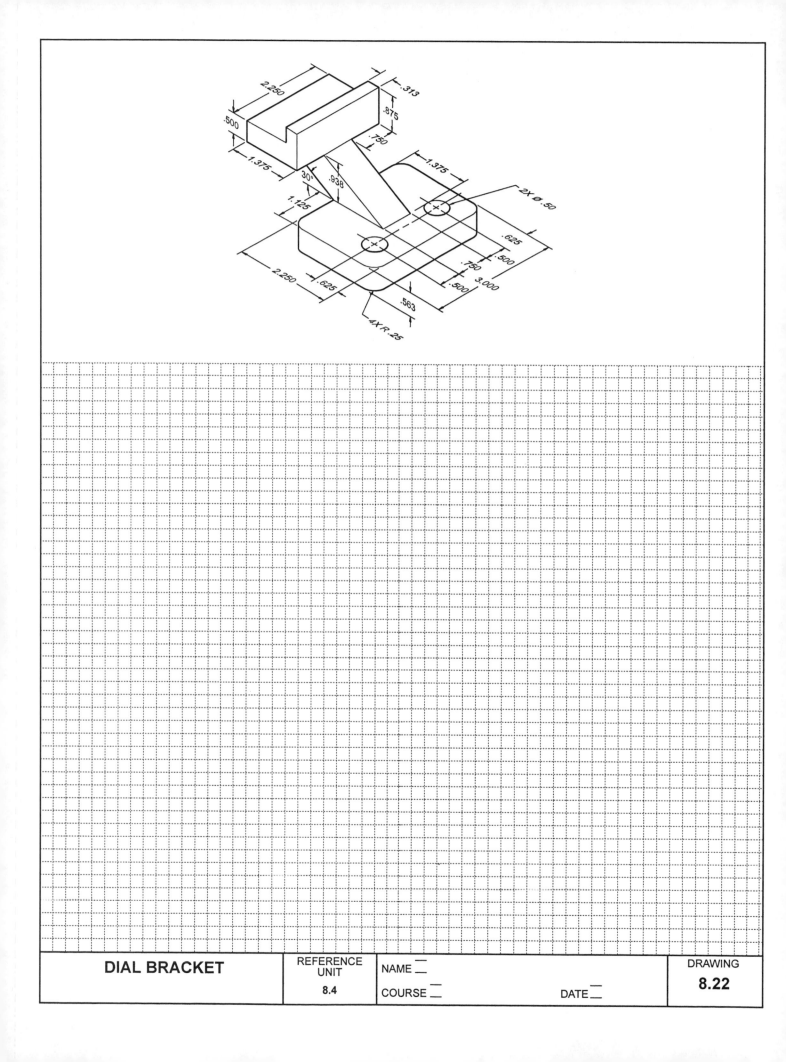

2.250
.313
.875
.500
.750
1.375
1.375
30°
.938
2X Ø .50
1.125
.625
.500
2.250
.750
.625
3.000
.500
.563
4X R .25

| DIAL BRACKET | REFERENCE UNIT | NAME __ | DRAWING |
| | 8.4 | COURSE __   DATE __ | 8.22 |

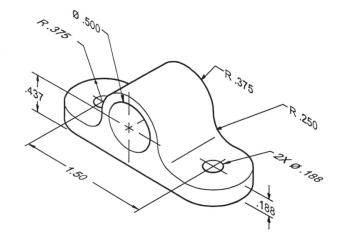

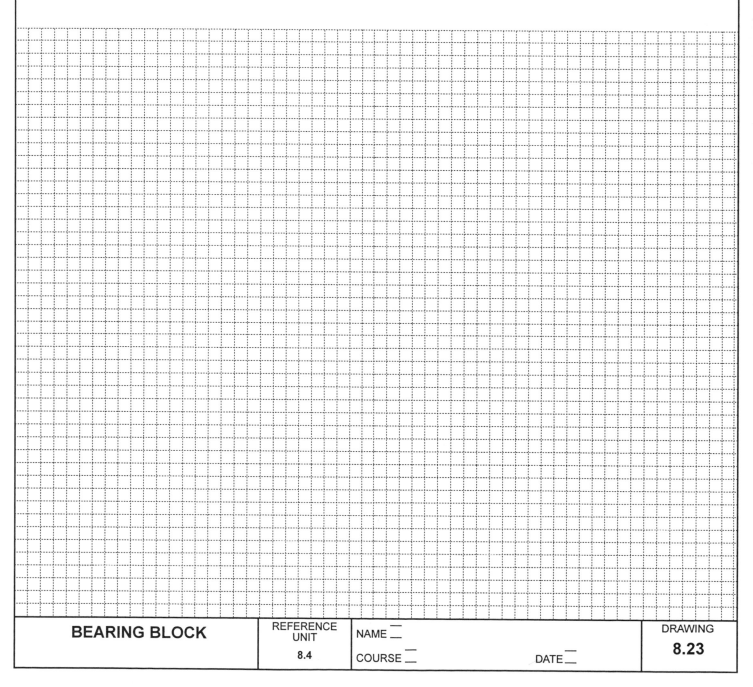

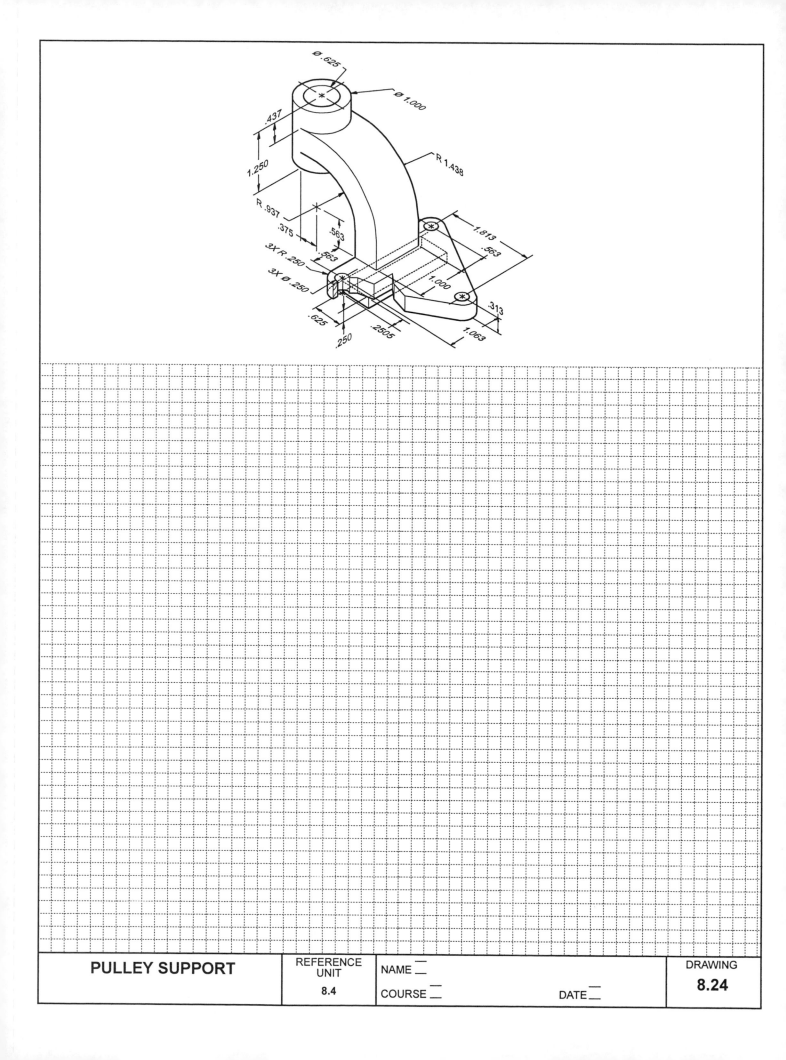

PULLEY SUPPORT

REFERENCE
UNIT

8.4

NAME __

COURSE __   DATE __

DRAWING

8.24

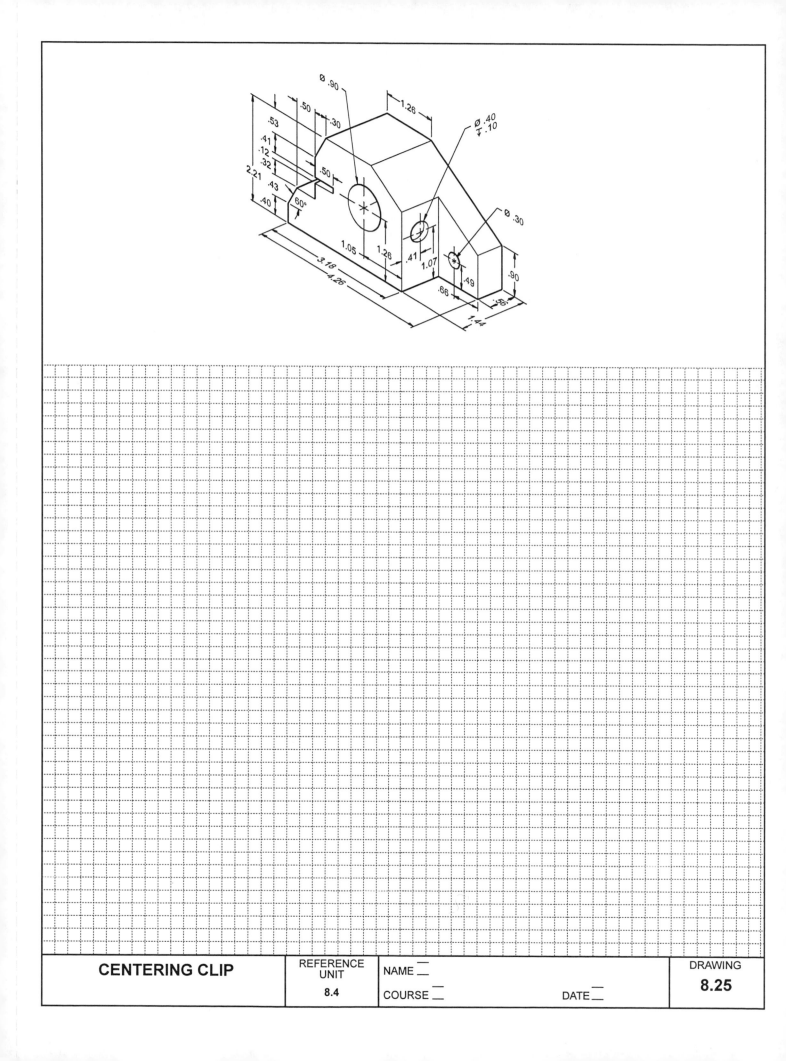

Ø .90

.50

.30

.53

.41

.12

.32

2.21

.43

.40

60°

.50

1.26

Ø .40
↧ .10

Ø .30

1.05

1.26

.41

1.07

3.18

4.26

.66

.49

.90

.56

1.44

| CENTERING CLIP | REFERENCE UNIT 8.4 | NAME — COURSE — DATE — | DRAWING 8.25 |

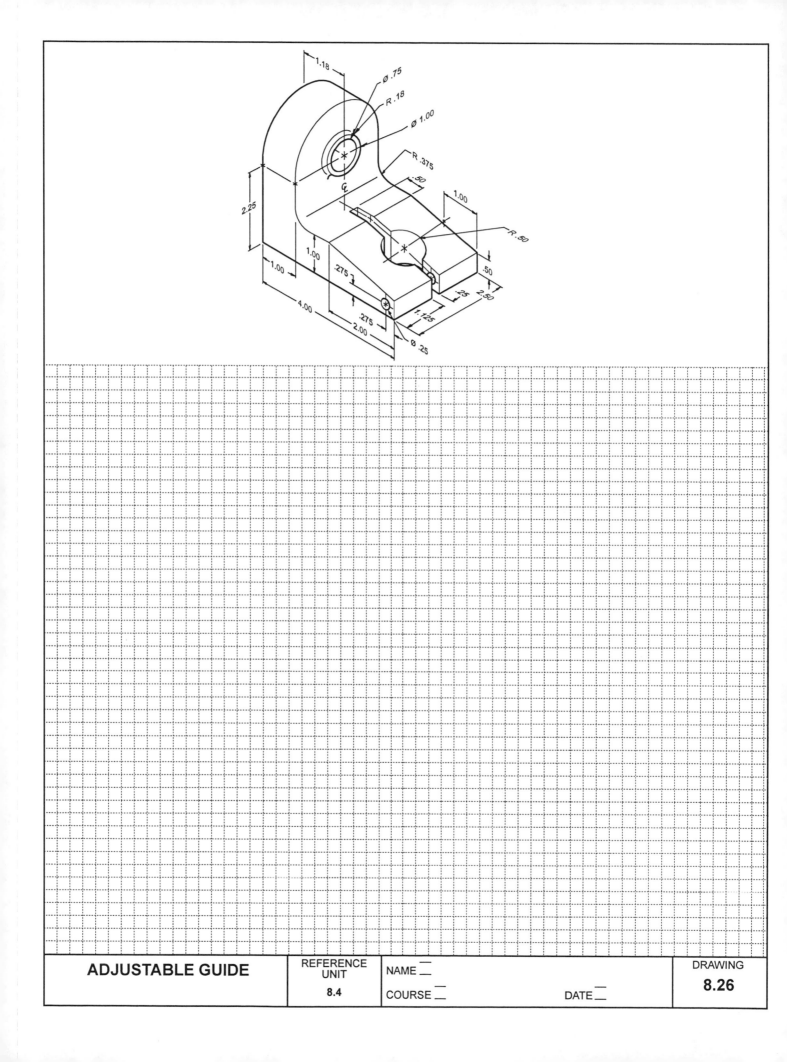

ADJUSTABLE GUIDE	REFERENCE UNIT	NAME __	DRAWING
	8.4	COURSE __ DATE __	8.26

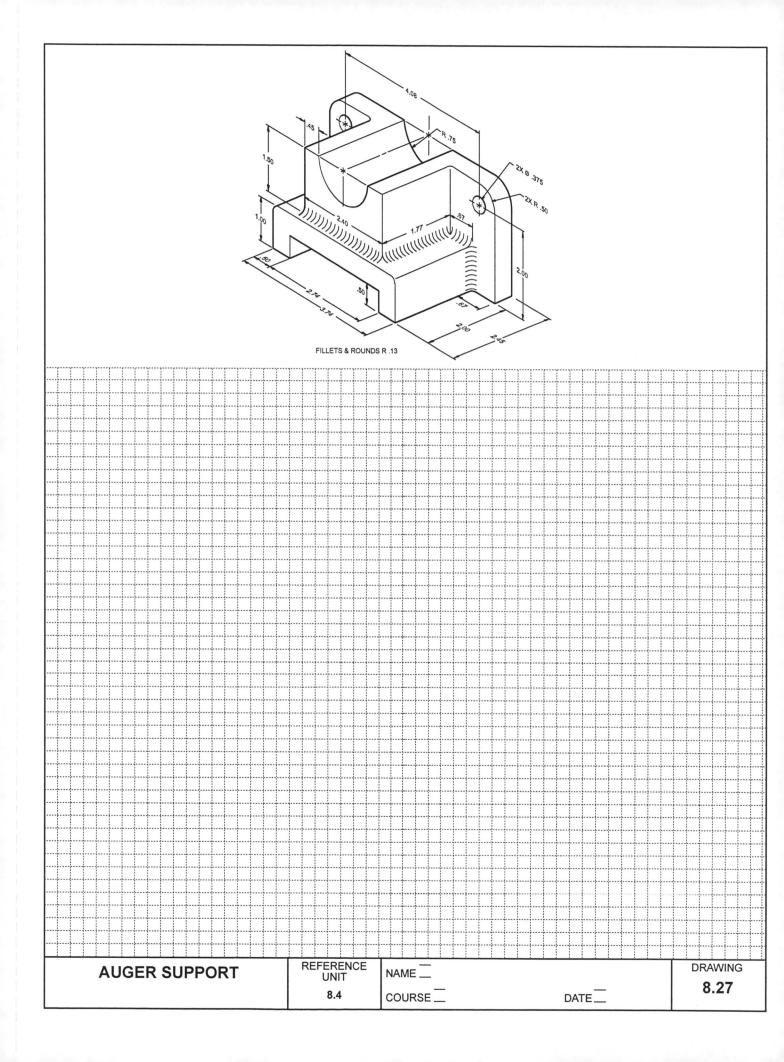

FILLETS & ROUNDS R .13

| AUGER SUPPORT | REFERENCE UNIT 8.4 | NAME __ COURSE __ DATE __ | DRAWING 8.27 |

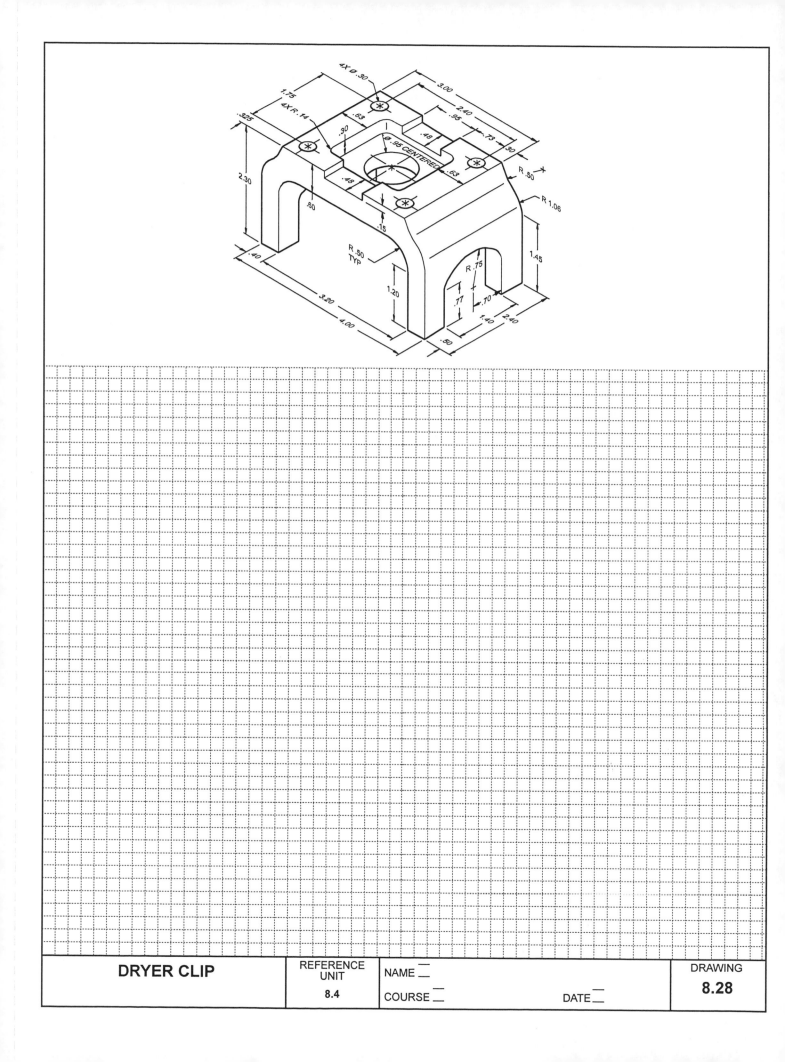

| DRYER CLIP | REFERENCE UNIT 8.4 | NAME __ COURSE __ DATE __ | DRAWING 8.28 |

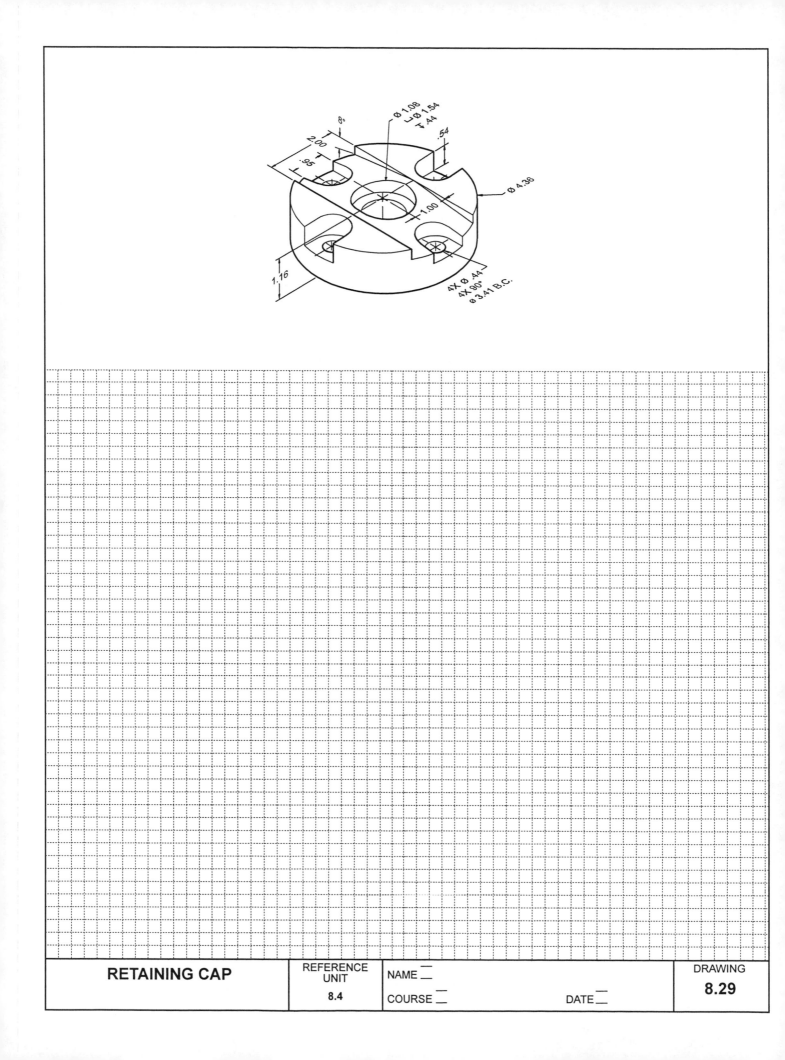

| RETAINING CAP | REFERENCE UNIT 8.4 | NAME __ COURSE __ DATE __ | DRAWING 8.29 |

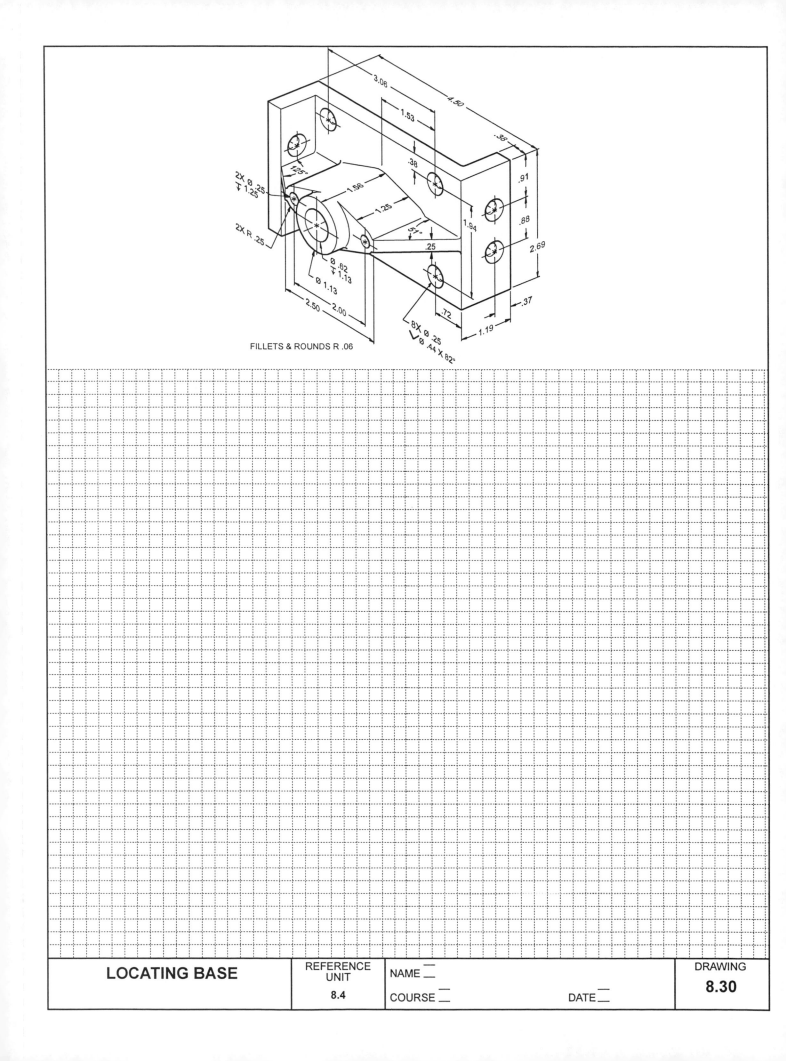

FILLETS & ROUNDS R .06

| LOCATING BASE | REFERENCE UNIT 8.4 | NAME — COURSE — | DATE — | DRAWING 8.30 |

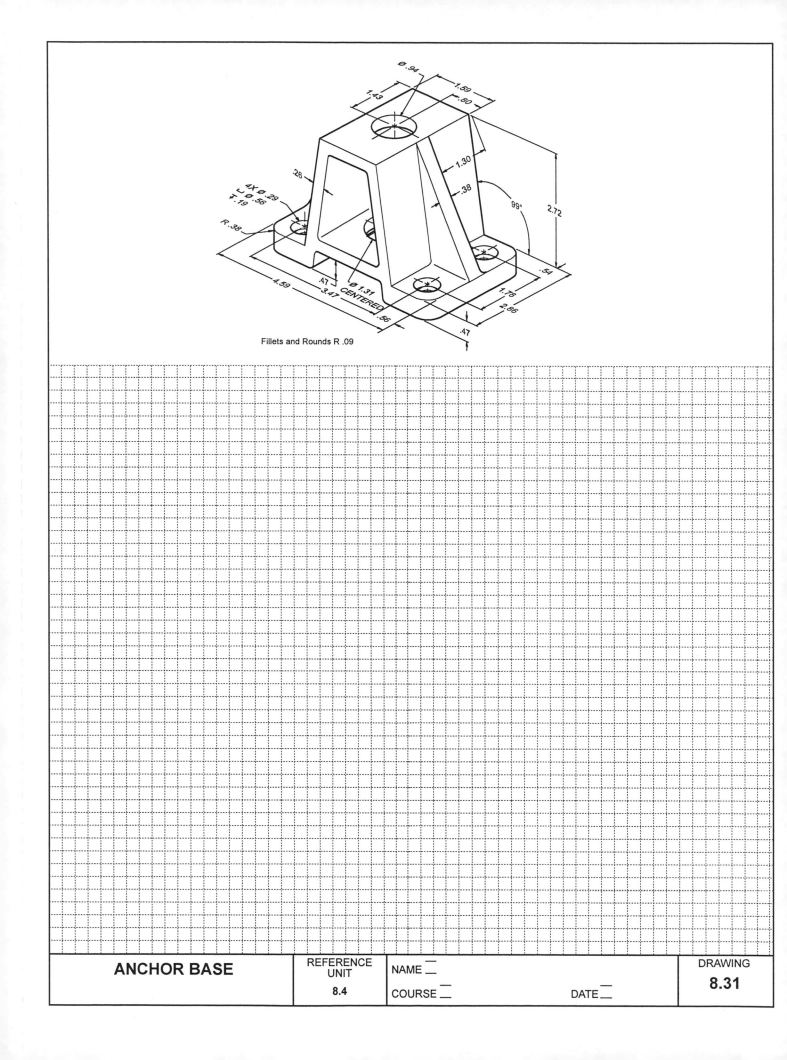

Fillets and Rounds R .09

ANCHOR BASE	REFERENCE UNIT	NAME __		DRAWING
	8.4	COURSE __	DATE __	8.31

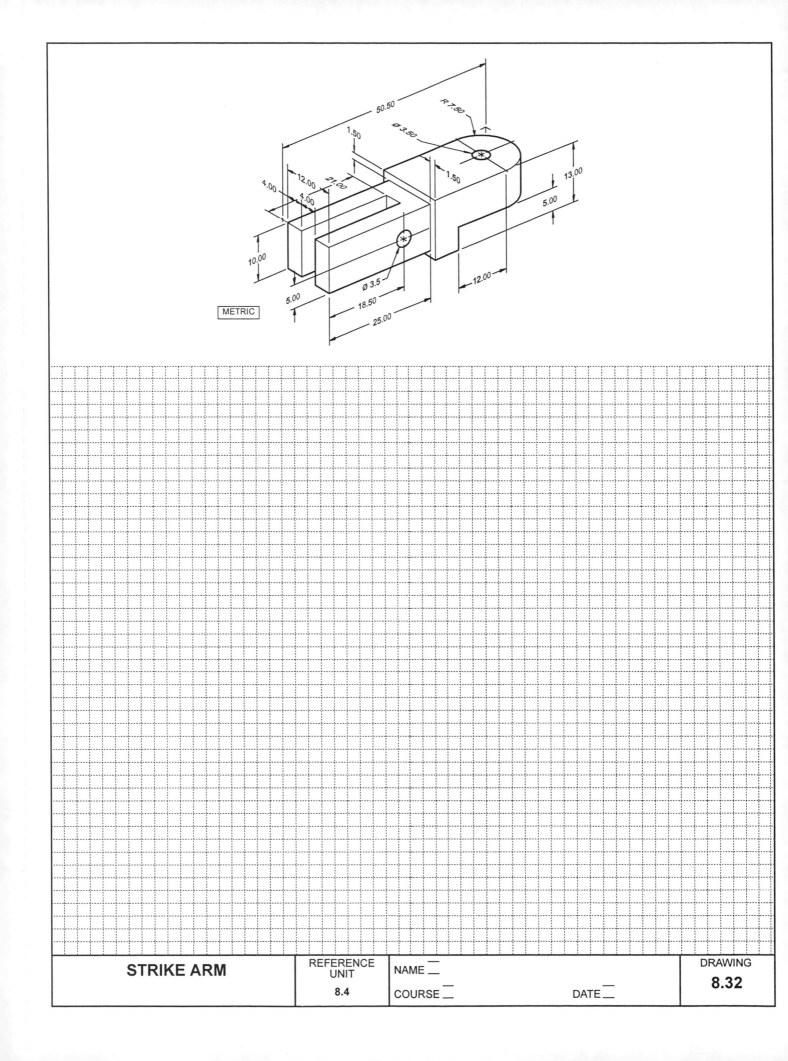

METRIC

STRIKE ARM	REFERENCE UNIT	NAME —		DRAWING
	8.4	COURSE —	DATE —	8.32

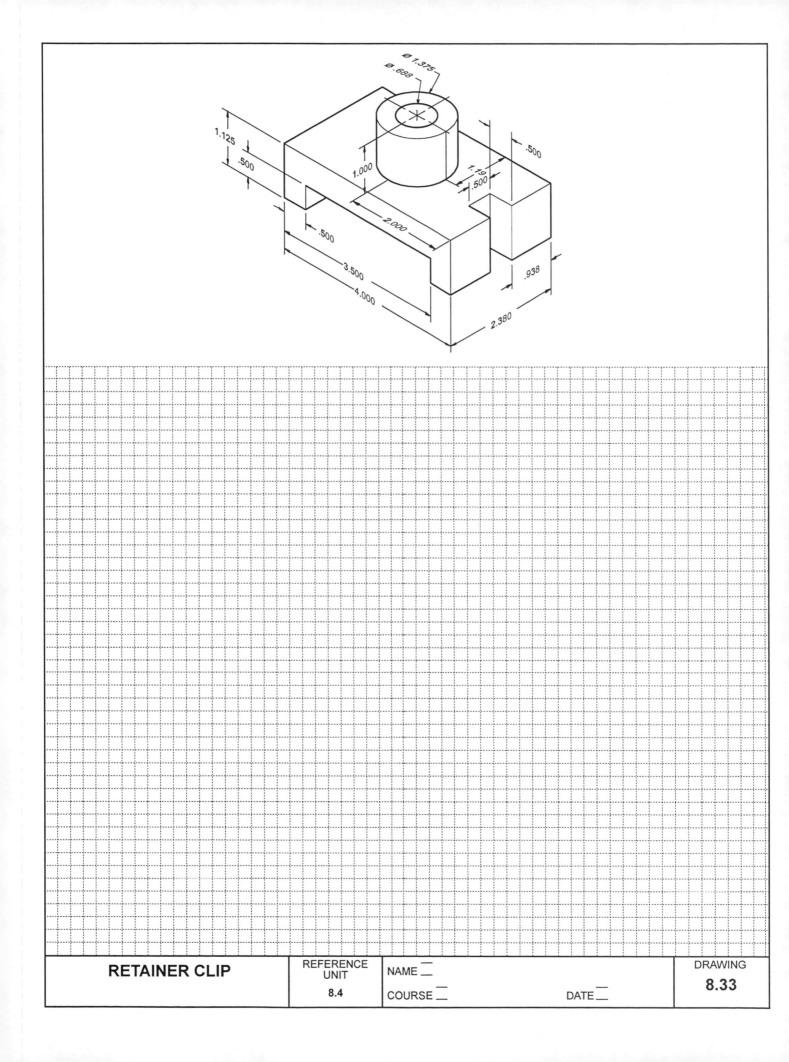

| RETAINER CLIP | REFERENCE UNIT 8.4 | NAME — COURSE — DATE — | DRAWING 8.33 |

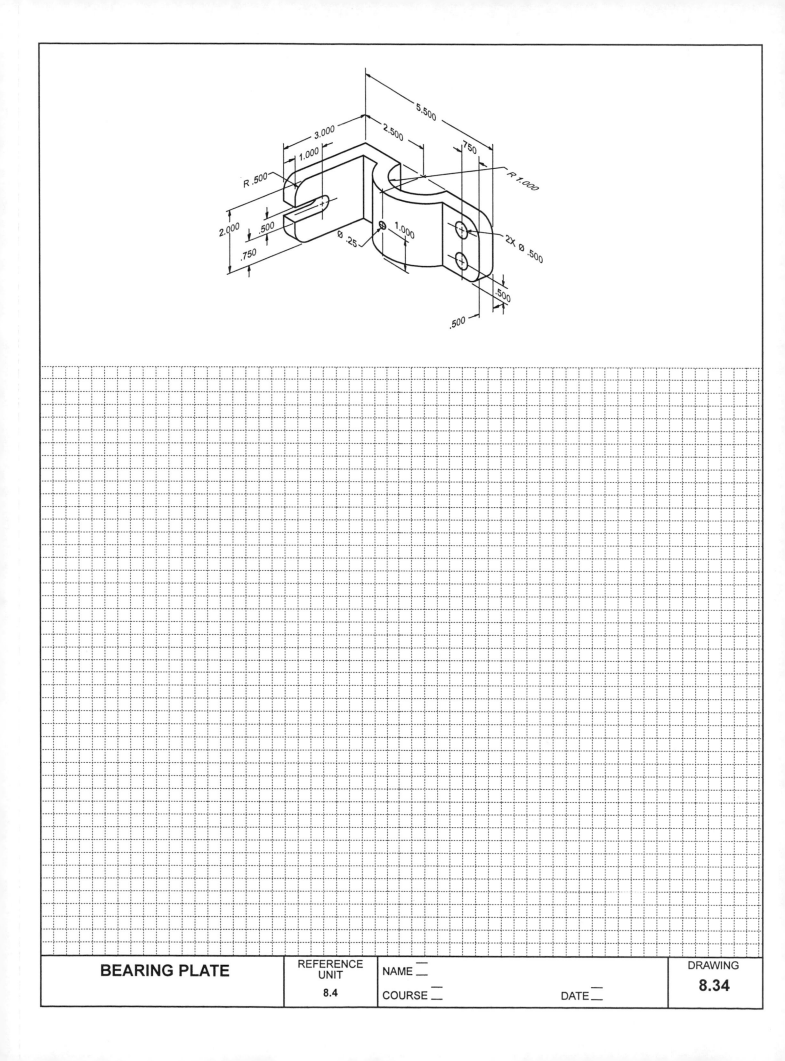

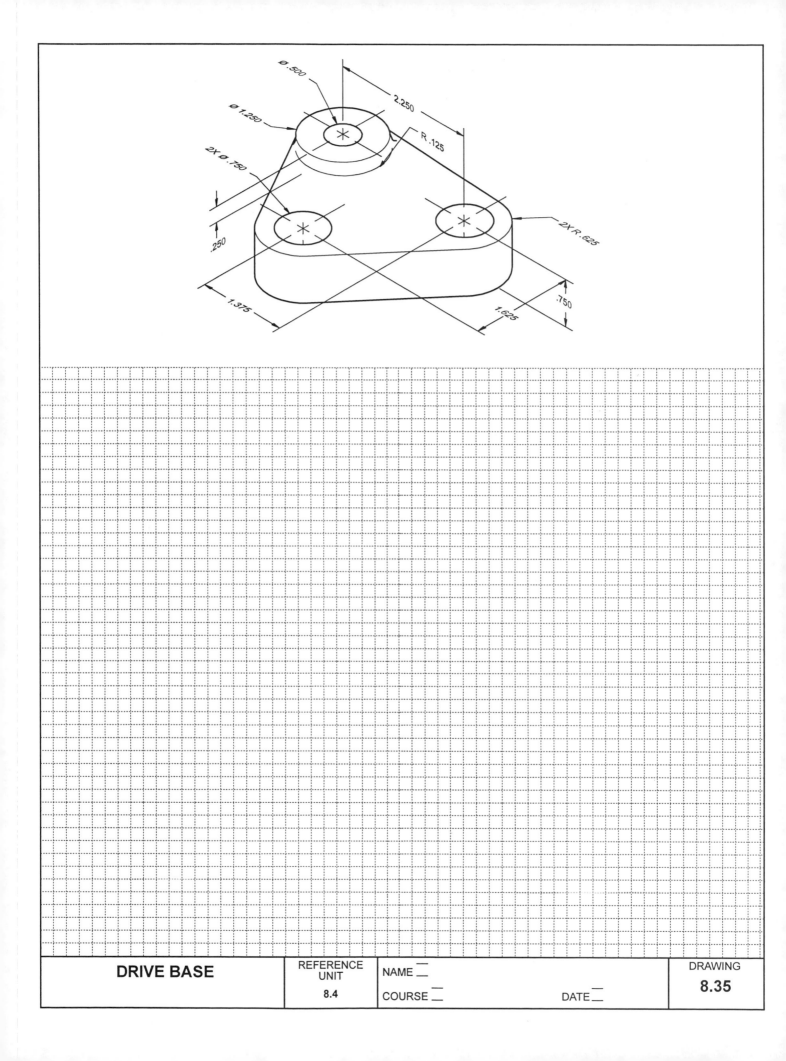

Ø .500

Ø 1.250

2X Ø .750

2.250

R .125

2X R .625

.250

.750

1.375

1.625

| DRIVE BASE | REFERENCE UNIT 8.4 | NAME ___ COURSE ___ DATE ___ | DRAWING 8.35 |

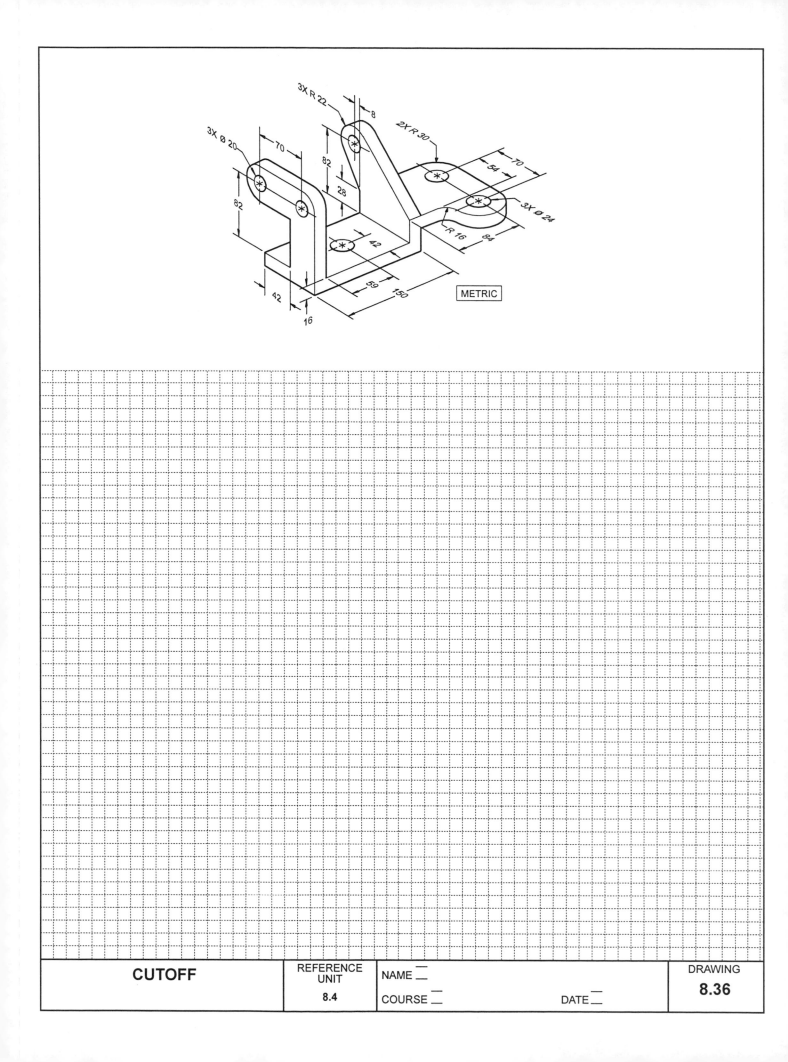

3X R 22

8

2X R 30

3X ø 20

70

82

82

28

54

70

3X ø 24

R 16

84

42

59

150

METRIC

42

16

CUTOFF

REFERENCE
UNIT

8.4

NAME __

COURSE __

DATE __

DRAWING

8.36

A.

B.

C.

D.

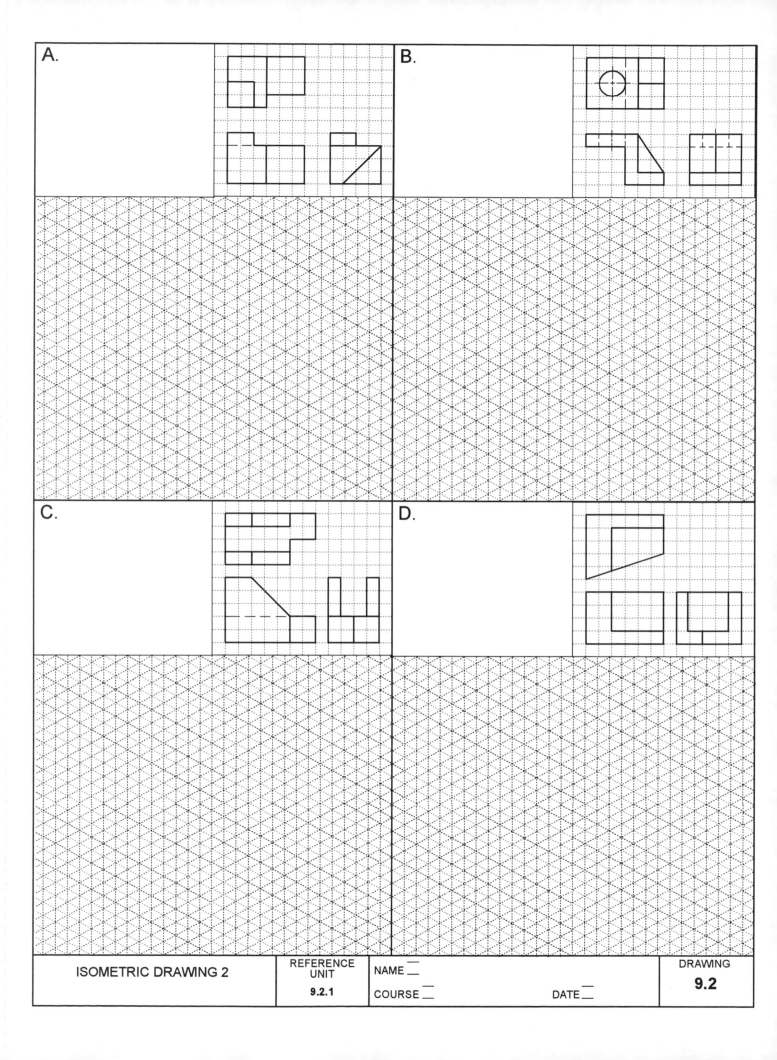

A.

B.

C.

D.

| ISOMETRIC DRAWING 2 | REFERENCE UNIT 9.2.1 | NAME __ COURSE __ DATE __ | DRAWING 9.2 |

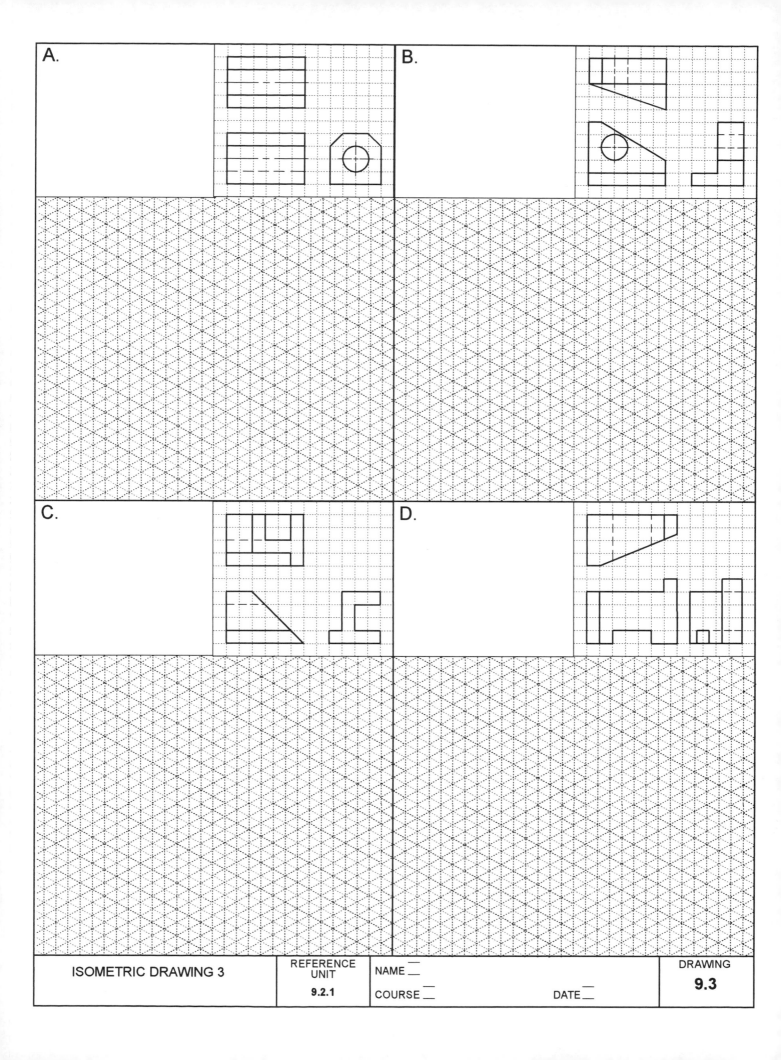

A.

B.

C.

D.

| ISOMETRIC DRAWING 3 | REFERENCE UNIT 9.2.1 | NAME __ COURSE __ DATE __ | DRAWING 9.3 |

A.

B.

C.

D.

ISOMETRIC DRAWING 4

REFERENCE UNIT

9.2.1

NAME __

COURSE __

DATE __

DRAWING

9.4

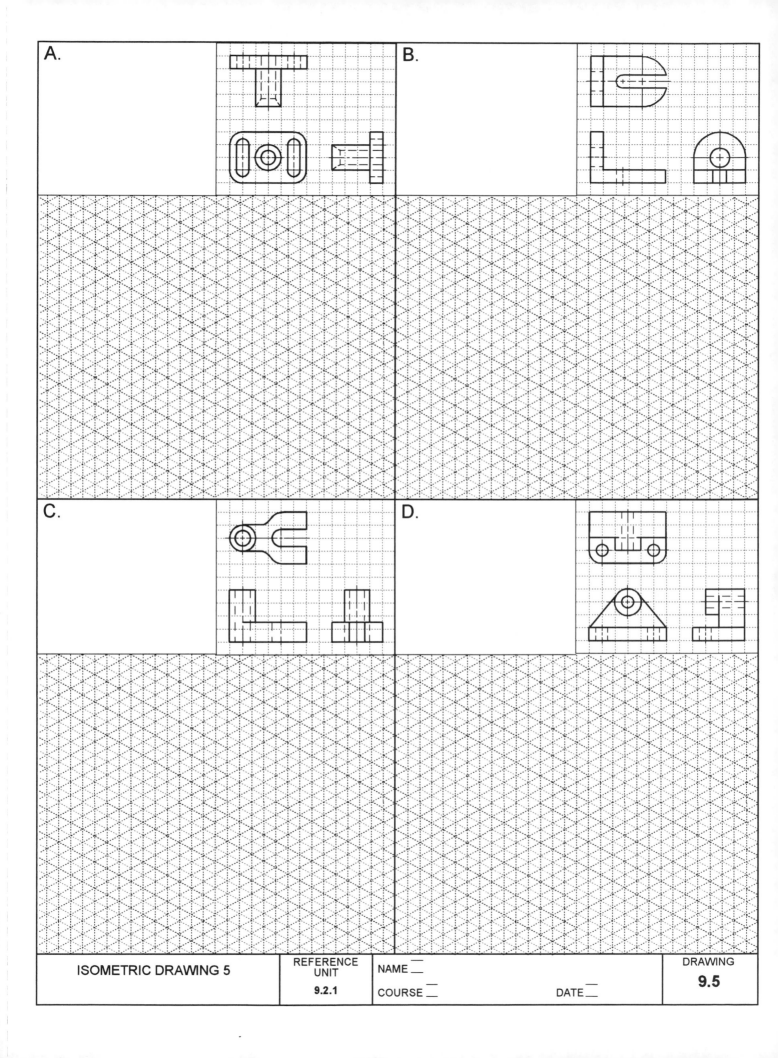

A.

B.

C.

D.

ISOMETRIC DRAWING 5

REFERENCE UNIT

9.2.1

NAME __

COURSE __

DATE __

DRAWING

9.5

A.

B.

C.

D.

| ISOMETRIC DRAWING 6 | REFERENCE UNIT 9.2.1 | NAME __ COURSE __ DATE __ | DRAWING 9.6 |

A.

B.

C.

D.

A.

B.

C.

D.

| OBLIQUE DRAWING 1 | REFERENCE UNIT 9.5 | NAME ___ COURSE ___ | DATE ___ | DRAWING 9.8 |

A.

B.

C.

D.

A.

B.

C.

D.

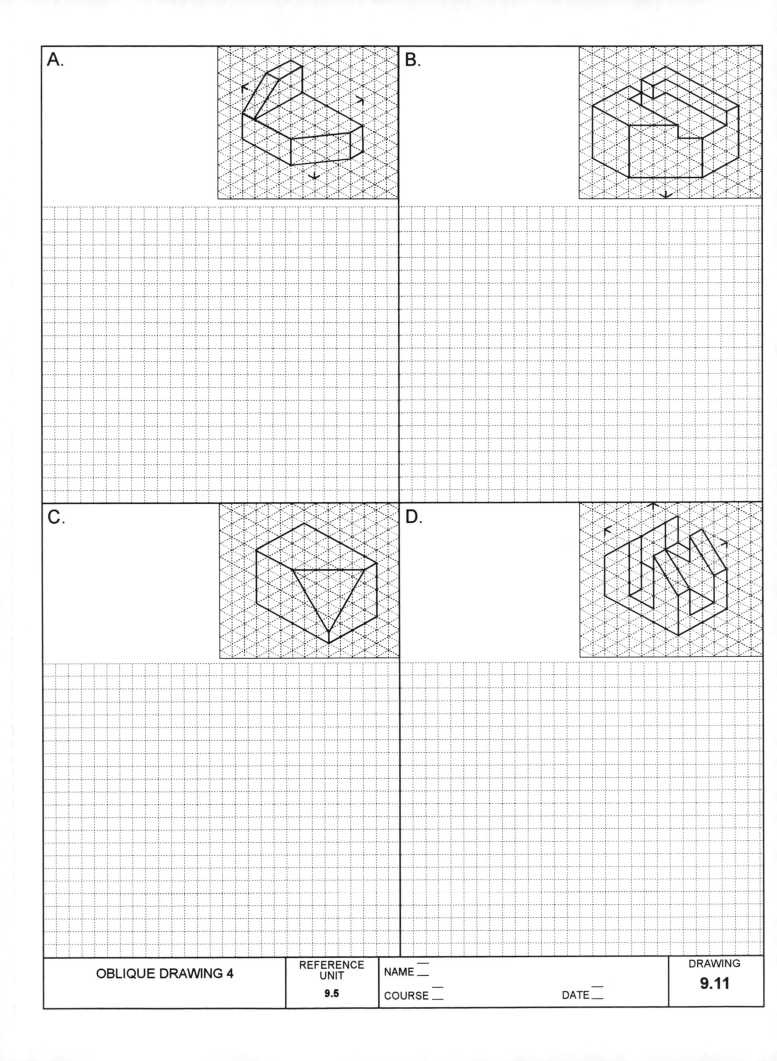

A.

B.

C.

D.

| OBLIQUE DRAWING 4 | REFERENCE UNIT 9.5 | NAME __
 COURSE __   DATE __ | DRAWING 9.11 |

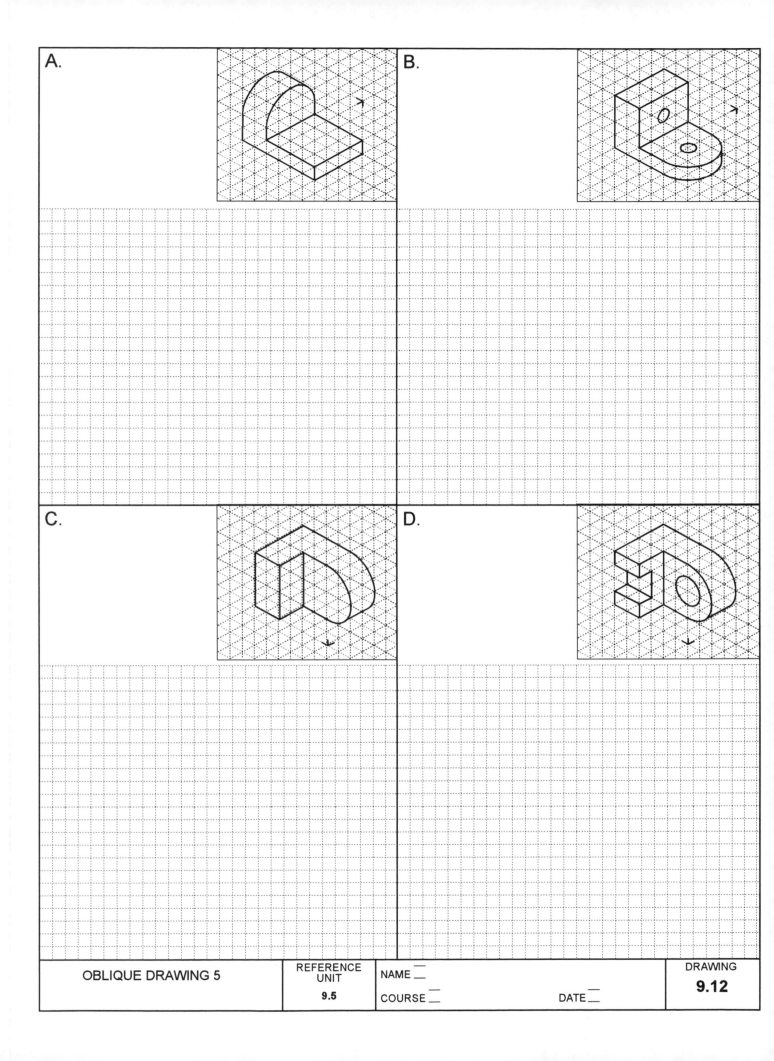

A.

B.

C.

D.

OBLIQUE DRAWING 5

REFERENCE UNIT

9.5

NAME __

COURSE __

DATE __

DRAWING

9.12

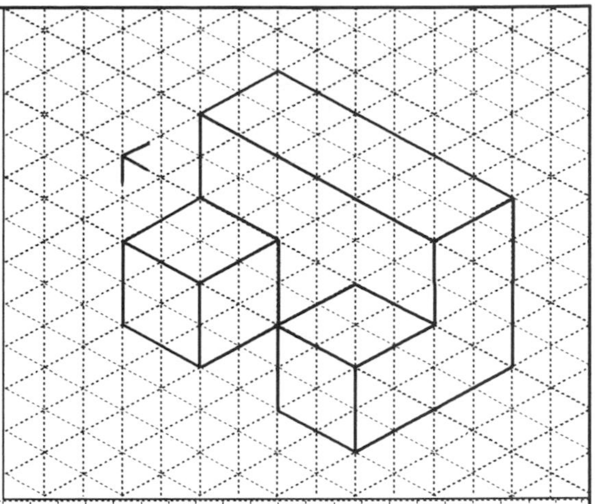

OBLIQUE DRAWING 6	REFERENCE UNIT 9.5	NAME ___ COURSE ___ DATE ___	DRAWING 9.13

| OBLIQUE DRAWING 7 | REFERENCE UNIT 9.5 | NAME __ COURSE __ | DATE __ | DRAWING 9.14 |

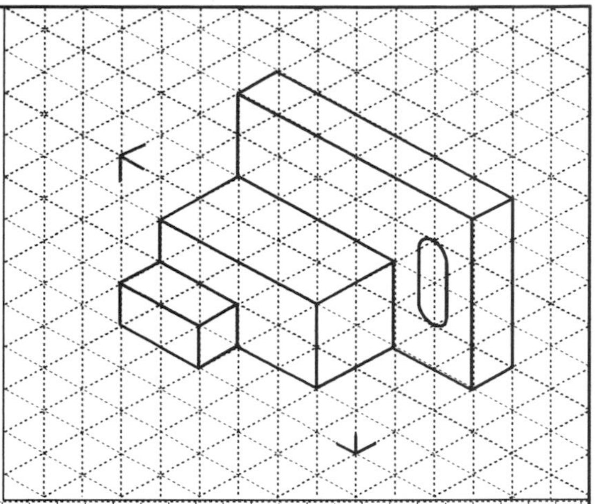

OBLIQUE DRAWING 8	REFERENCE UNIT 9.5	NAME ___ COURSE ___ DATE ___	DRAWING 9.15

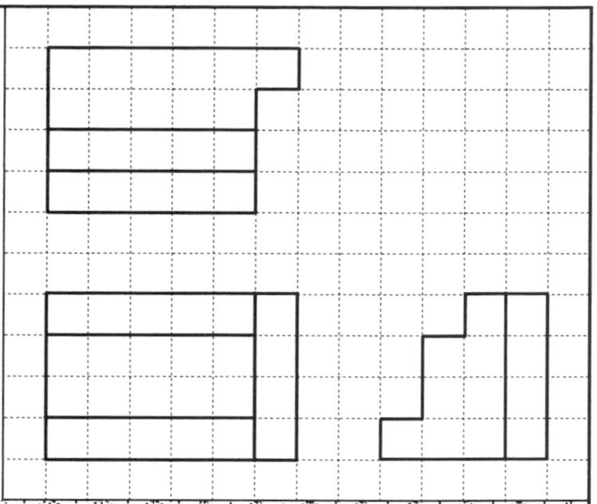

ISOMETRIC DRAWING 7	REFERENCE UNIT 9.2.1	NAME ___ COURSE ___ DATE ___	DRAWING 9.16

| ISOMETRIC DRAWING 9 | REFERENCE UNIT | NAME __ | | DRAWING |
| | 9.2.1 | COURSE __ | DATE __ | **9.18** |

ISOMETRIC DRAWING 10 | REFERENCE UNIT **9.2.1** | NAME __ COURSE __ DATE __ | DRAWING **9.19**

| ISOMETRIC DRAWING 11 | REFERENCE UNIT 9.2.1 | NAME __ COURSE __ DATE __ | DRAWING 9.20 |

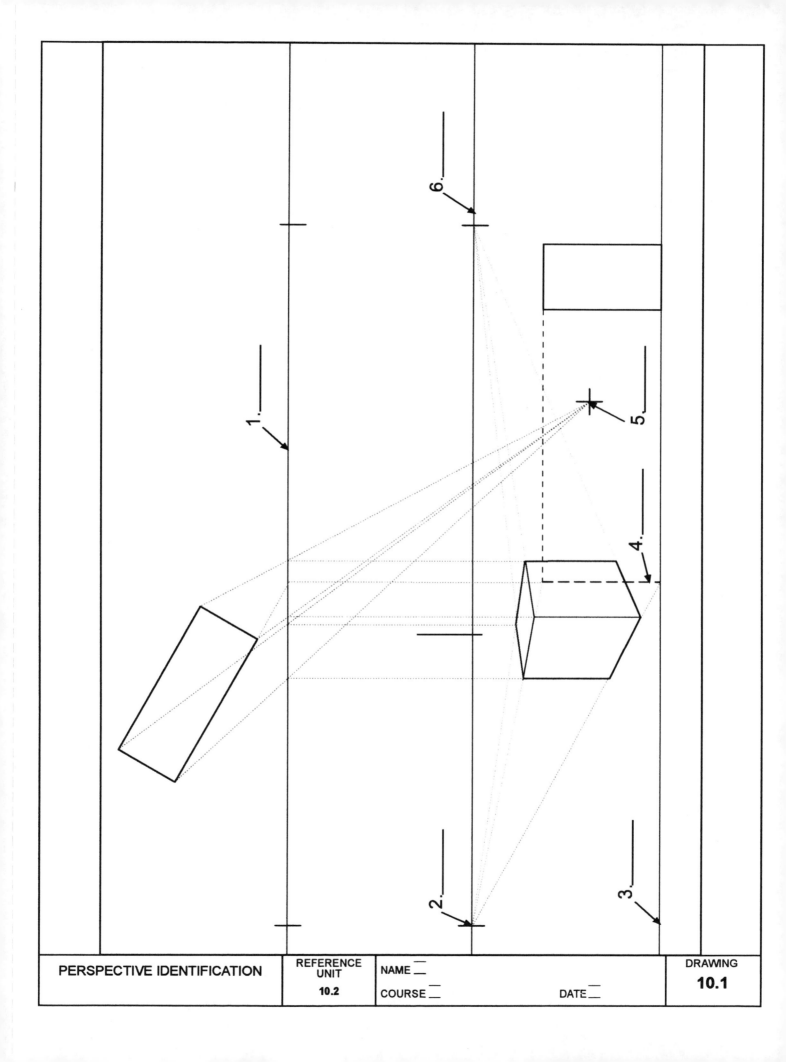

PERSPECTIVE IDENTIFICATION | REFERENCE UNIT 10.2 | NAME __ COURSE __ DATE __ | DRAWING 10.1

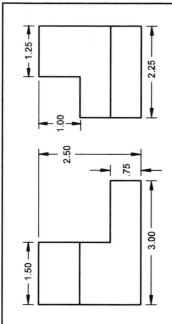

1.25
2.25
1.00

2.50
.75
1.50
3.00

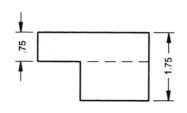

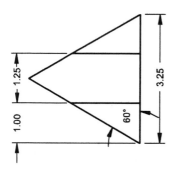

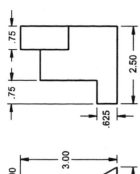

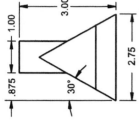

ONE-POINT PERSPECTIVE 3

REFERENCE UNIT

10.5

NAME __

COURSE __

DATE __

DRAWING

10.4

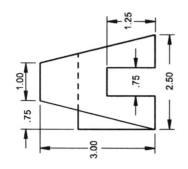

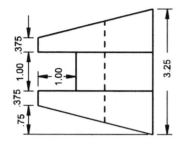

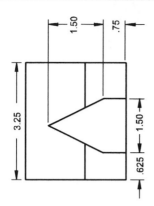

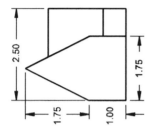

| TWO-POINT PERSPECTIVE 1 | REFERENCE UNIT 10.6 | NAME __ COURSE __ DATE __ | DRAWING 10.6 |

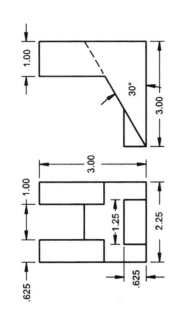

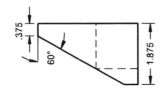

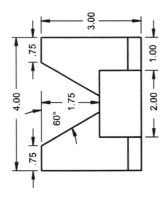

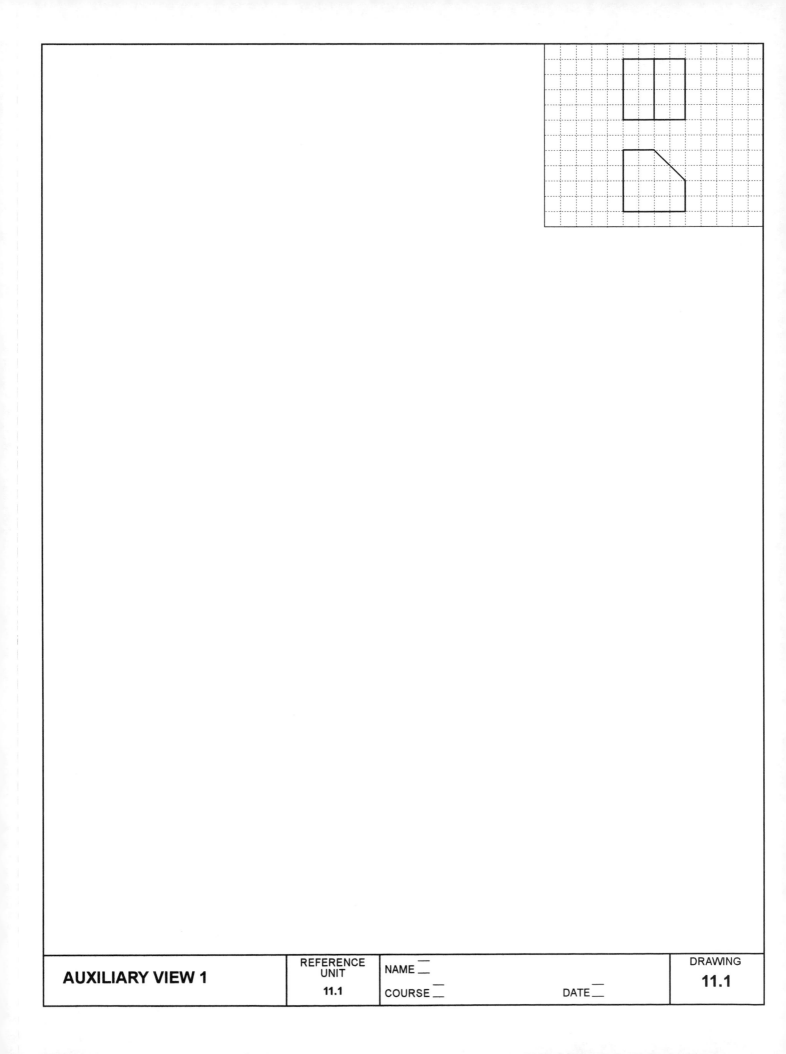

AUXILIARY VIEW 1

REFERENCE
UNIT

11.1

NAME __

COURSE __

DATE __

DRAWING

11.1

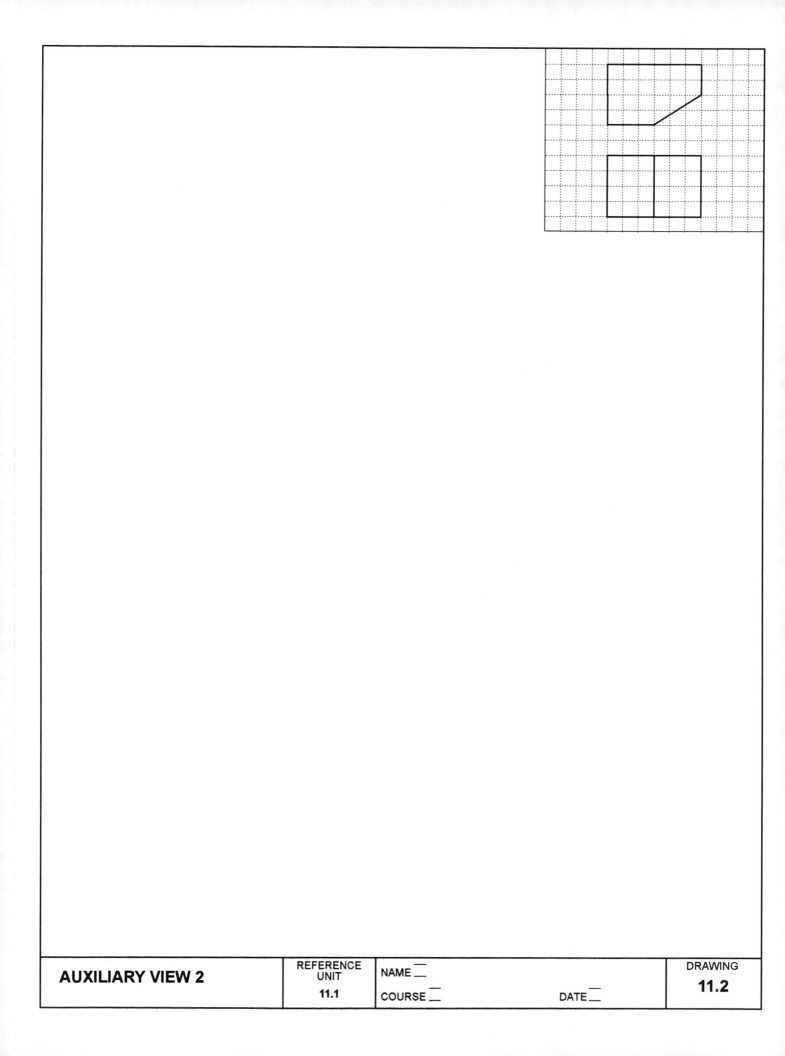

AUXILIARY VIEW 2

REFERENCE
UNIT

11.1

NAME ___

COURSE ___

DATE ___

DRAWING

11.2

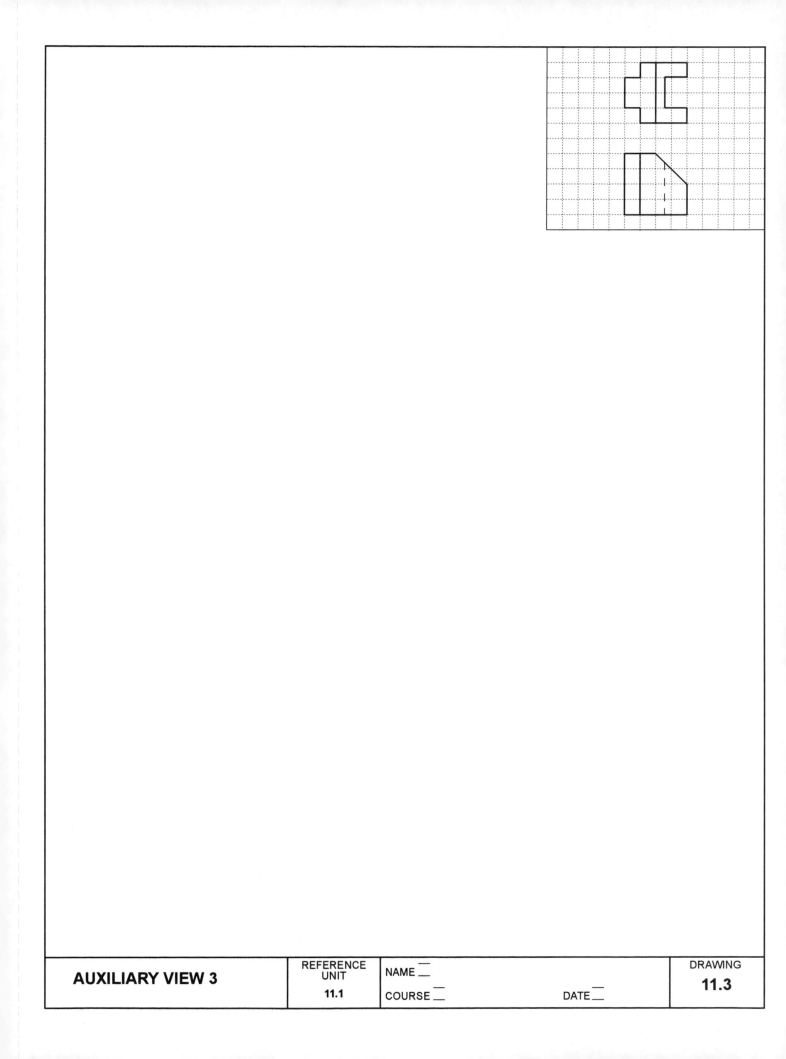

AUXILIARY VIEW 3

REFERENCE
UNIT

11.1

NAME __

COURSE __

DATE __

DRAWING

11.3

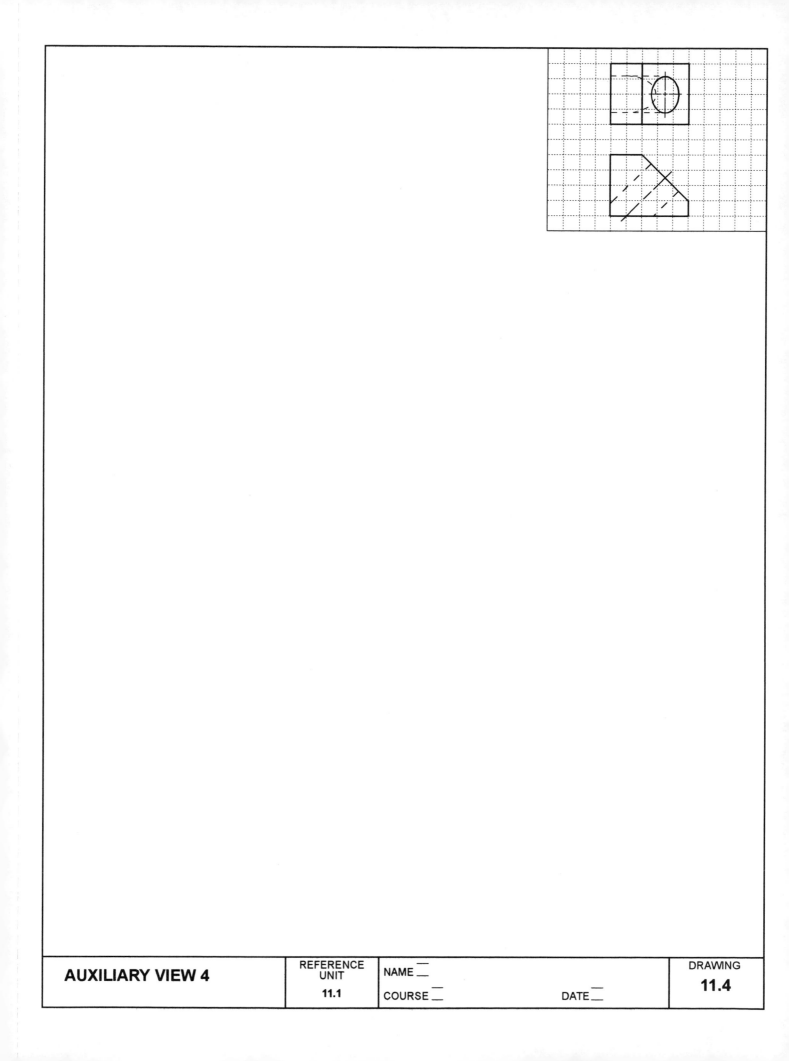

AUXILIARY VIEW 4

REFERENCE
UNIT

11.1

NAME __

COURSE __

DATE __

DRAWING

11.4

AUXILIARY VIEW 5

REFERENCE
UNIT

11.1

NAME __

COURSE __

DATE __

DRAWING

11.5

AUXILIARY VIEW 6

REFERENCE
UNIT

11.1

NAME —

COURSE —

DATE —

DRAWING

11.6

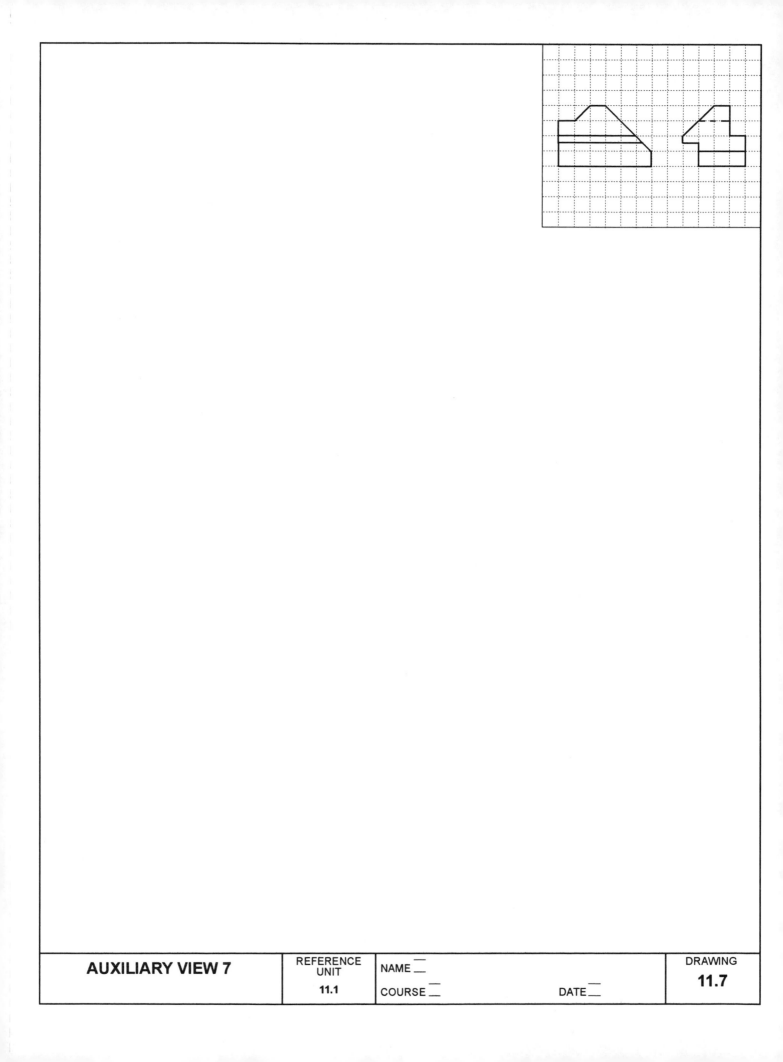

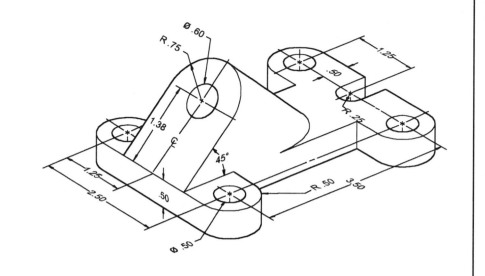

AUXILIARY VIEW 8

REFERENCE UNIT

11.1, 2, 3

NAME —

COURSE —

DATE —

DRAWING

11.8

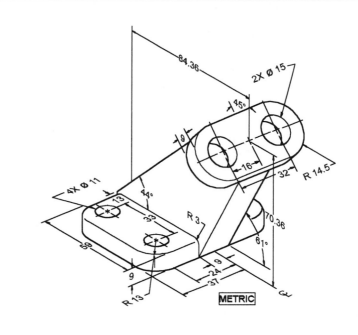

AUXILIARY VIEW 9

REFERENCE UNIT

11.1, 2, 3

NAME __

COURSE __

DATE __

DRAWING

11.9

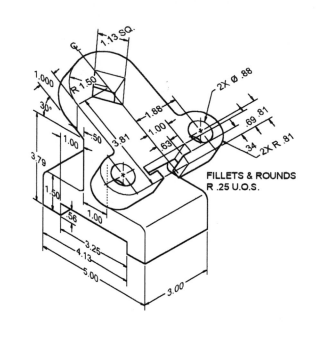

FILLETS & ROUNDS
R .25 U.O.S.

AUXILIARY VIEW 10

REFERENCE UNIT

11.1, 2, 3

NAME __

COURSE __

DATE __

DRAWING

11.10

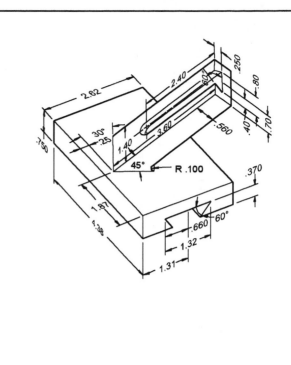

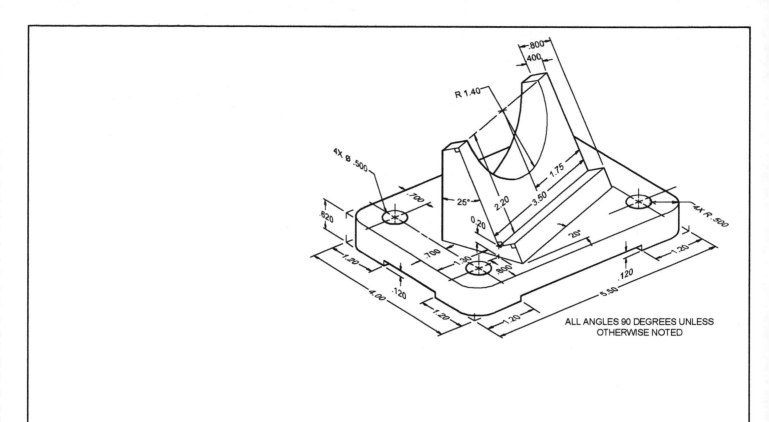

ALL ANGLES 90 DEGREES UNLESS
OTHERWISE NOTED

AUXILIARY VIEW 12

REFERENCE
UNIT

11.1, 2, 3

NAME __

COURSE __   DATE __

DRAWING

11.12

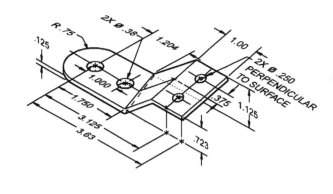

AUXILIARY VIEW 13

REFERENCE
UNIT

11.1, 2, 3

NAME __

COURSE __

DATE __

DRAWING

11.13

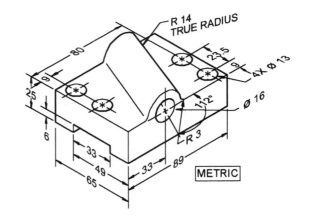

R 14
TRUE RADIUS

80

23.5

9

4X Ø 13

25

9

112°

Ø 16

6

R 3

33

33

89

49

METRIC

65

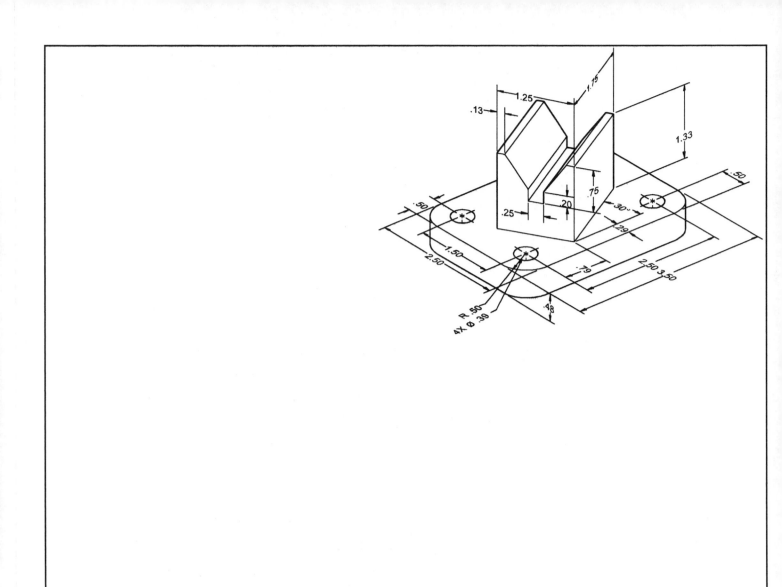

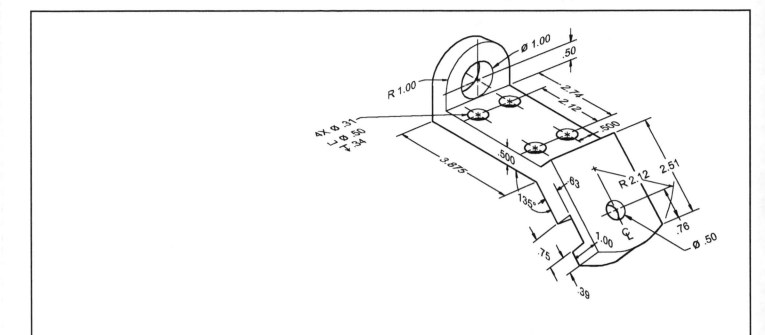

Ø 1.00
.50
R 1.00
2.74
2.12
4X Ø .31
⌴ Ø .50
↧ .34
.500
3.875
.500
.63
2.51
135°
R 2.12
.76
.75
1.00
℄
Ø .50
.39

| AUXILIARY VIEW 16 | REFERENCE UNIT 11.1, 2, 3 | NAME __ COURSE __ DATE __ | DRAWING 11.16 |

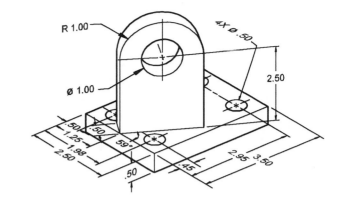

AUXILIARY VIEW 17

REFERENCE UNIT

11.1, 2, 3

NAME __

COURSE __

DATE __

DRAWING

11.17

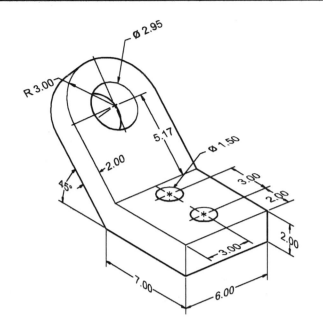

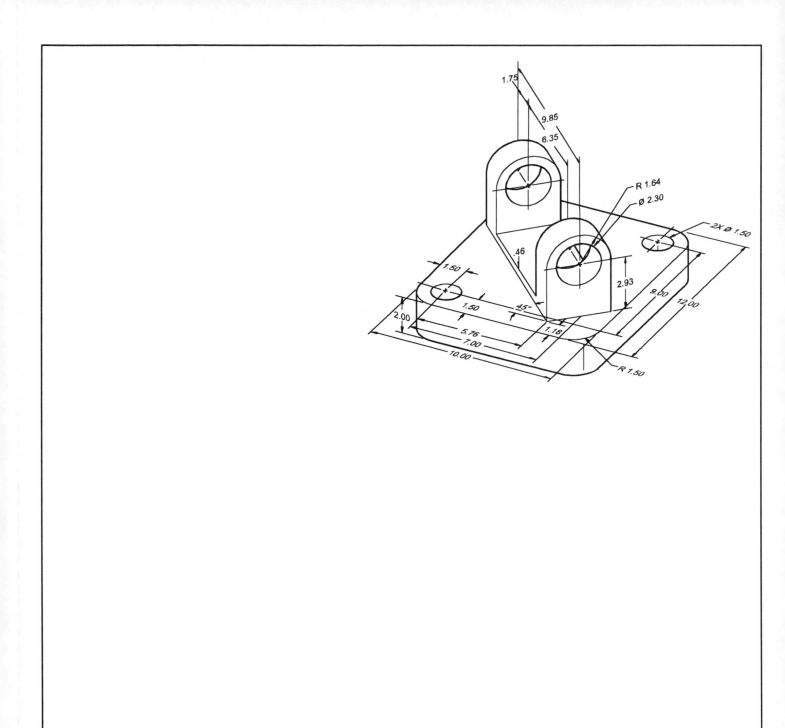

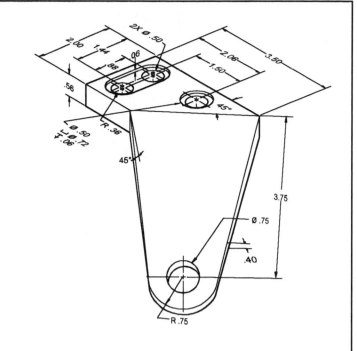

2X Ø .50
2.00
1.44
.88
.06
2.06
3.50
1.50
.56
45°
R .36
Ø .50
Ø .72
45°
3.75
Ø .75
.40
R .75

AUXILIARY VIEW 20	REFERENCE UNIT	NAME __		DRAWING
	11.1, 2, 3	COURSE __	DATE __	**11.20**

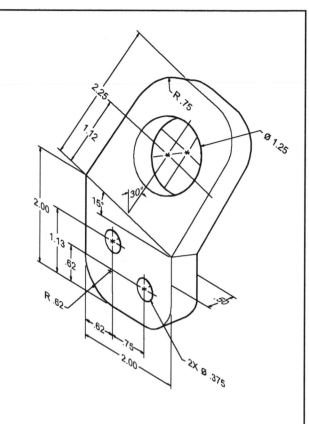

AUXILIARY VIEW 21

REFERENCE
UNIT

11.1, 2, 3

NAME __

COURSE __

DATE __

DRAWING

11.21

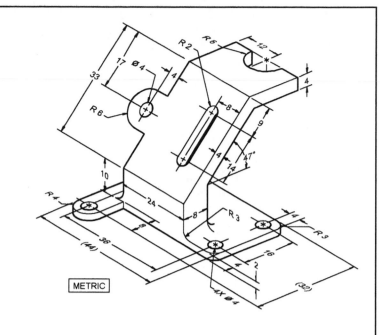

METRIC

AUXILIARY VIEW 22

REFERENCE UNIT

11.1, 2, 3

NAME __

COURSE __

DATE __

DRAWING

11.22

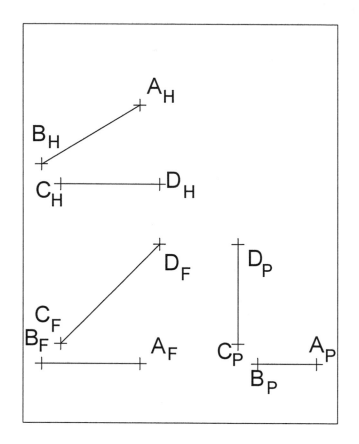

AB_____

CD_____

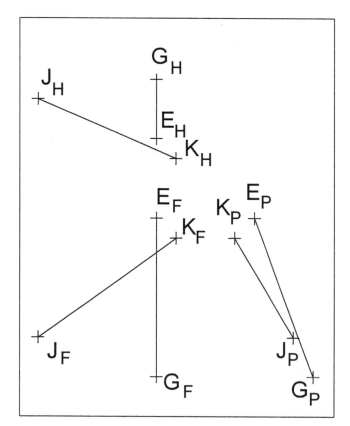

EG_____

JK _____

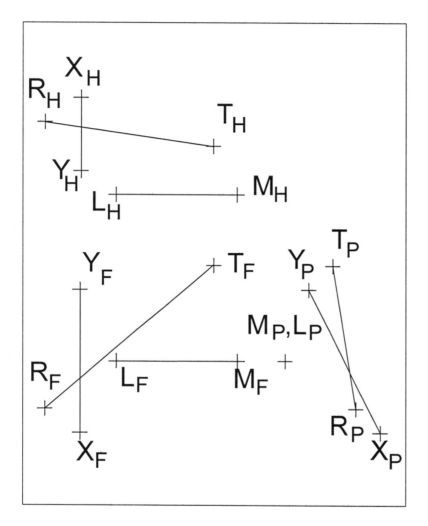

A _____

B _____

C _____

D _____

E _____

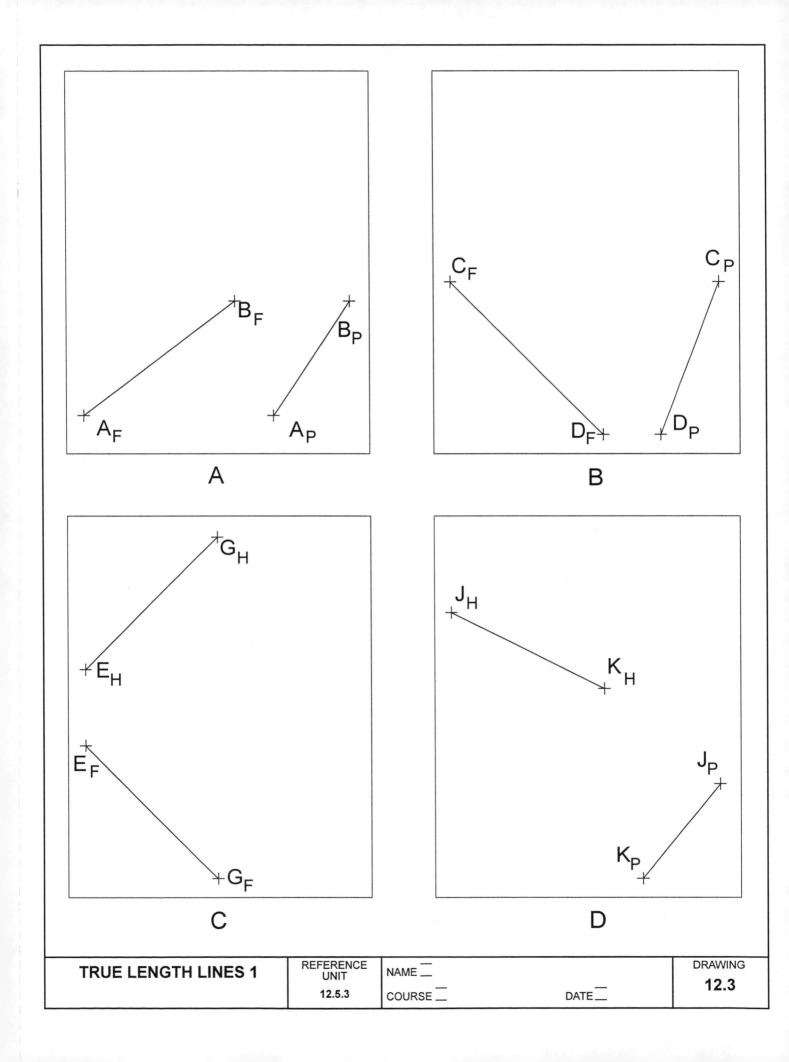

A

B

C

D

TRUE LENGTH LINES 1

REFERENCE
UNIT

12.5.3

NAME

COURSE

DATE

DRAWING

12.3

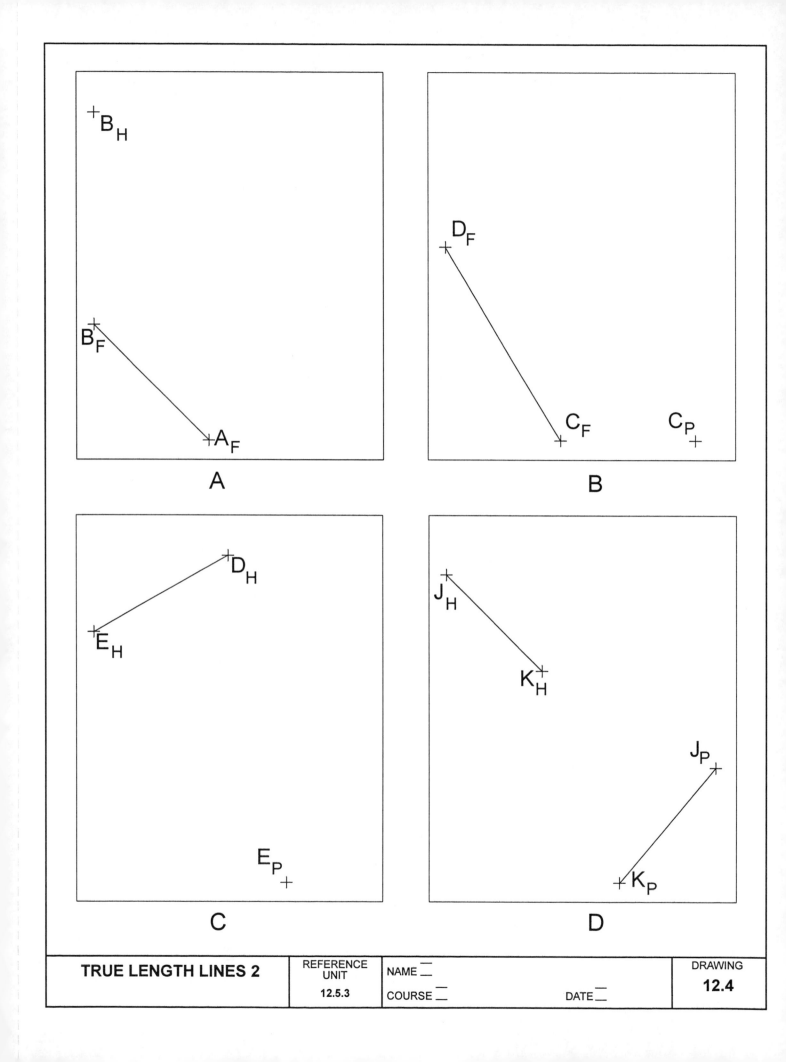

A

B

C

D

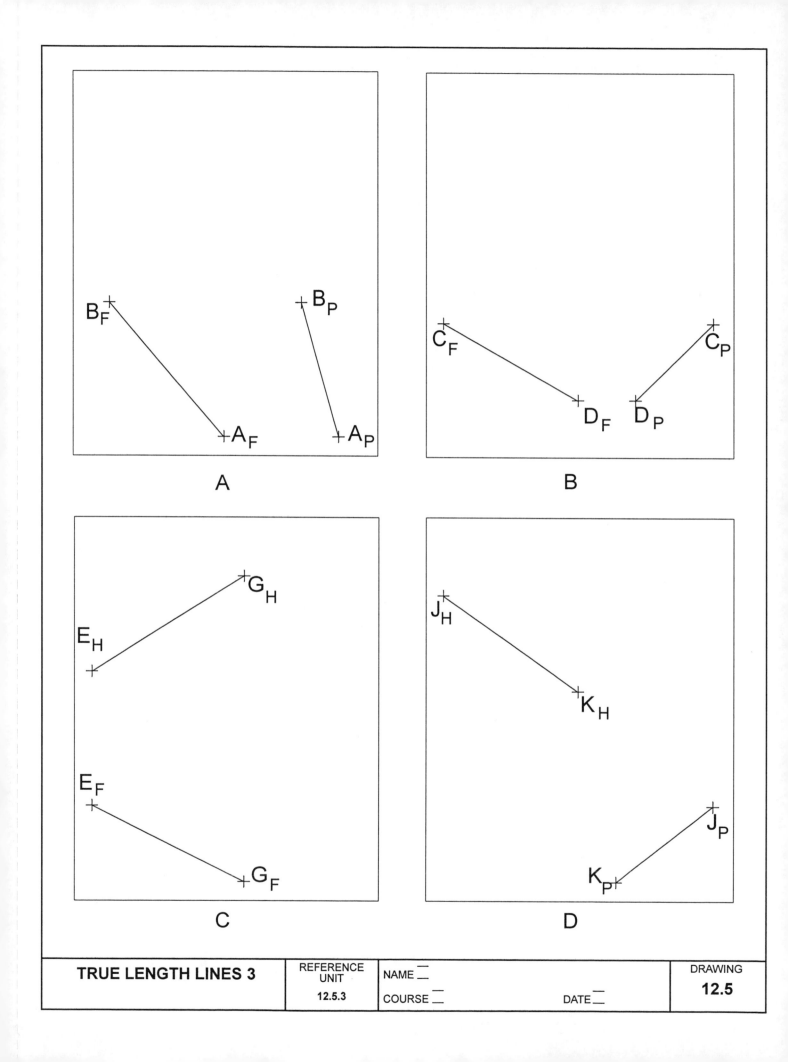

A

B

C

D

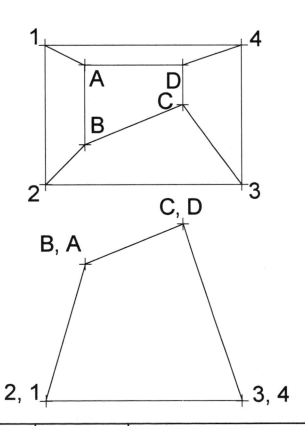

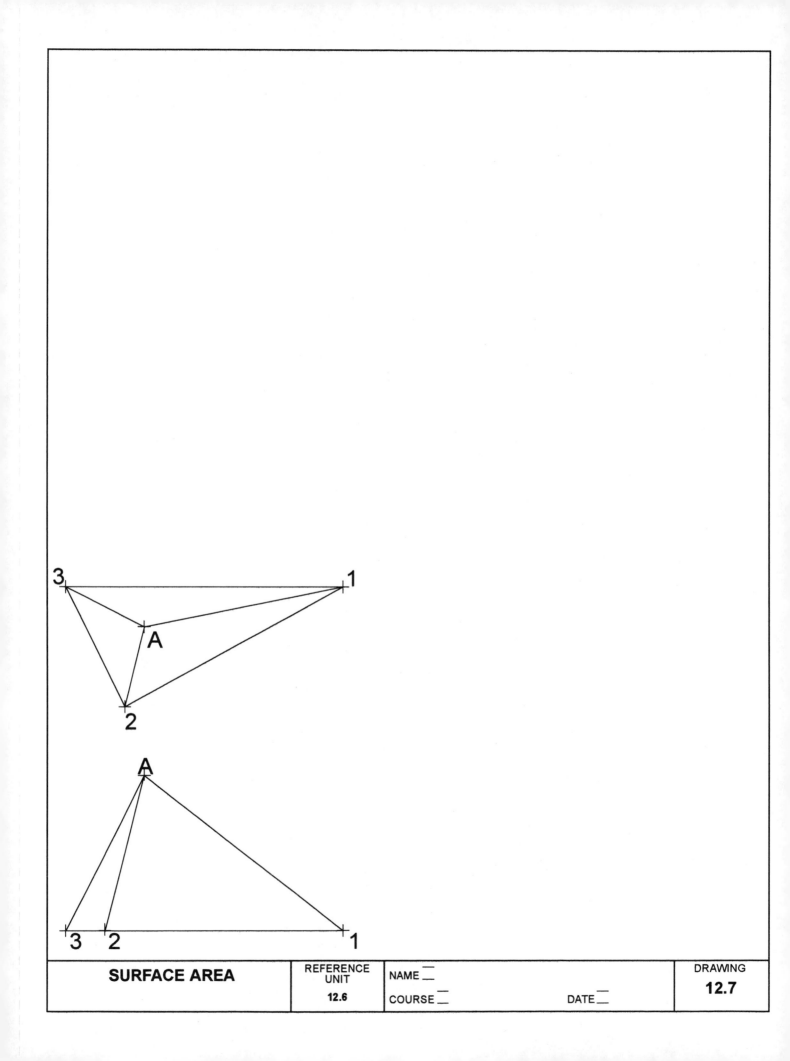

| SURFACE AREA | REFERENCE UNIT 12.6 | NAME __ COURSE __ DATE __ | DRAWING 12.7 |

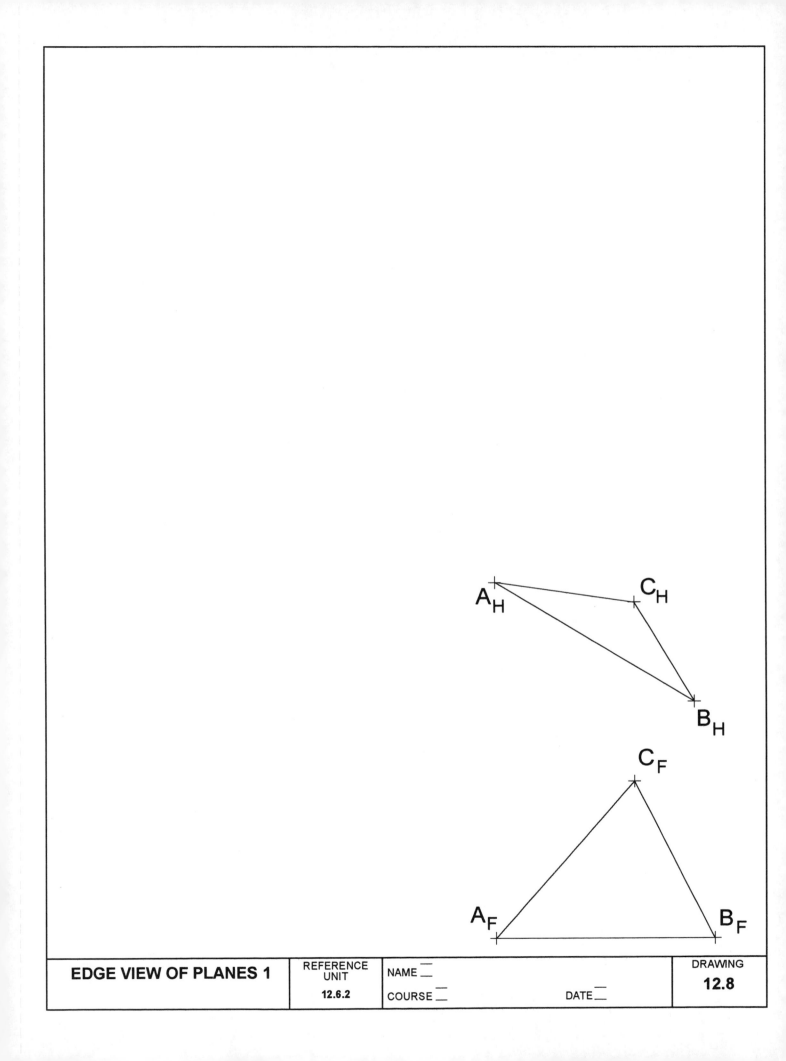

E_F
$+$

E_P
$+$

D_F
$+$

D_P
$+$

F_P
$+$

$+F_F$

EDGE VIEW OF PLANES 2	REFERENCE UNIT	NAME __		DRAWING
	12.6.2	COURSE __	DATE __	**12.9**

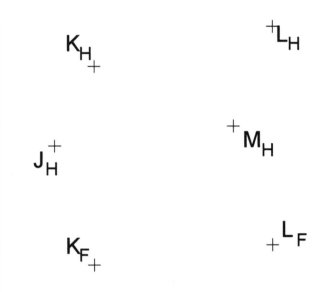

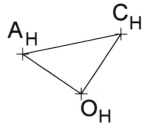

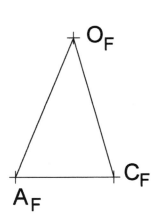

| TRUE SIZE OF PLANES 1 | REFERENCE UNIT 12.6.3 | NAME __ COURSE __ | DATE __ | DRAWING 12.11 |

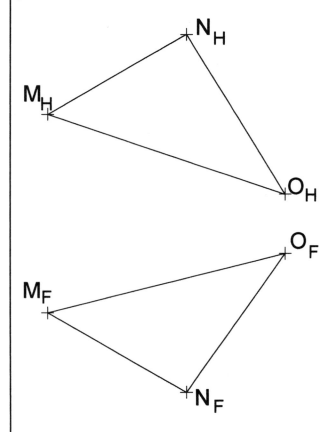

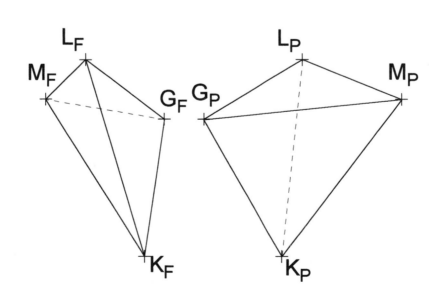

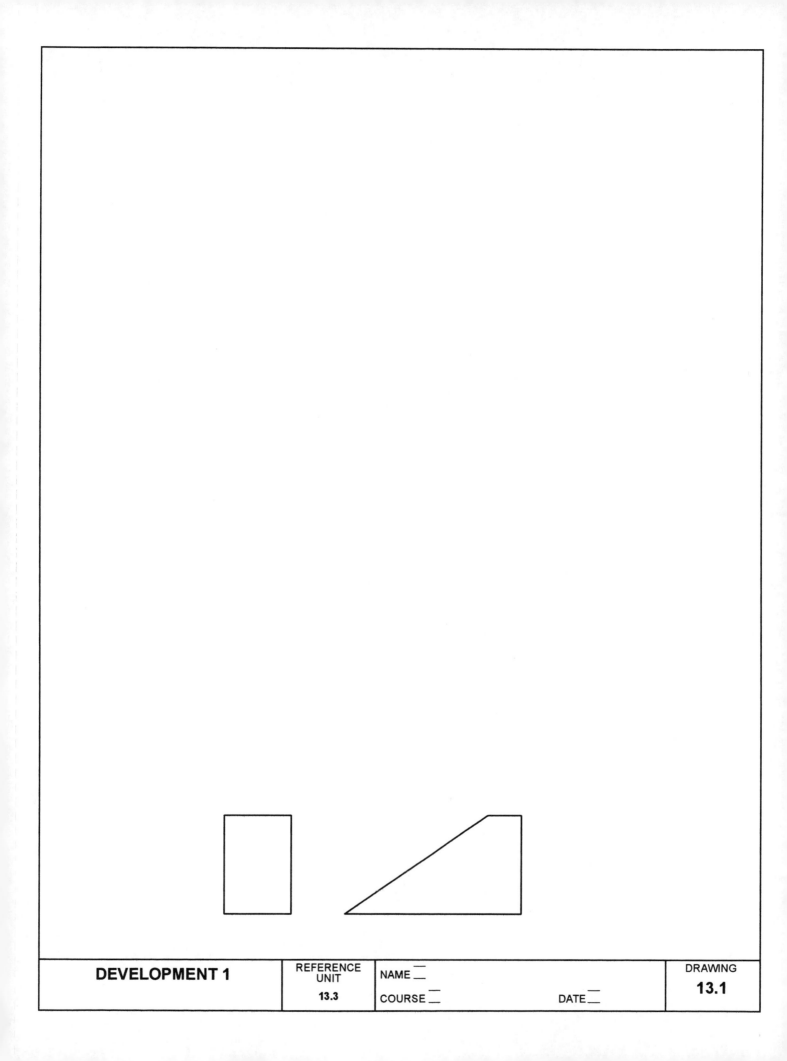

DEVELOPMENT 1

REFERENCE
UNIT

13.3

NAME __

COURSE __

DATE __

DRAWING

13.1

DEVELOPMENT 2

REFERENCE
UNIT

13.3

NAME __

COURSE __

DATE __

DRAWING

13.2

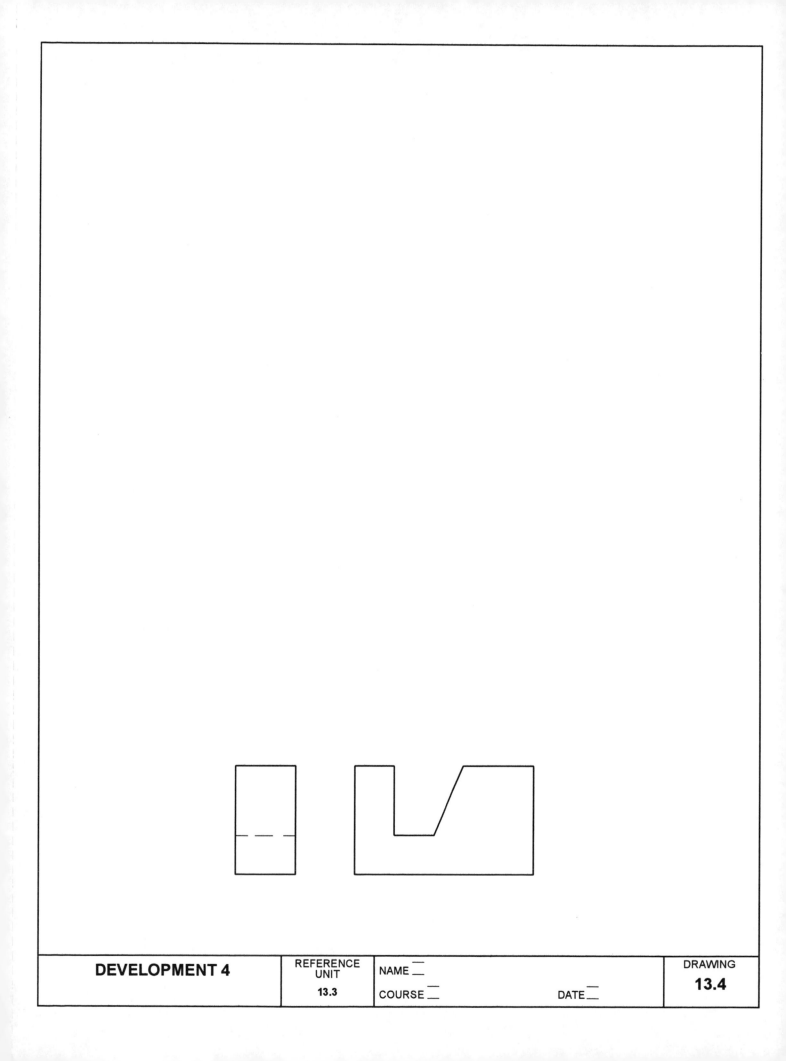

DEVELOPMENT 4

REFERENCE
UNIT

13.3

NAME ___

COURSE ___

DATE ___

DRAWING

13.4

DEVELOPMENT 6

REFERENCE
UNIT

13.3

NAME —

COURSE —

DATE —

DRAWING

13.6

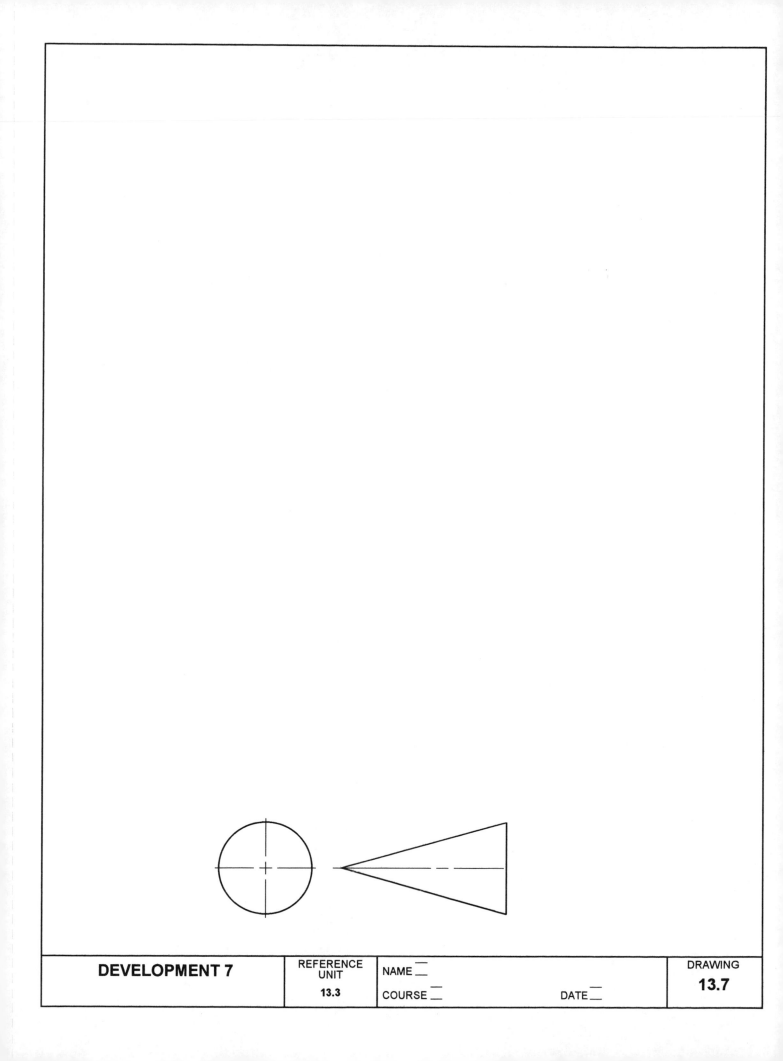

DEVELOPMENT 7

REFERENCE
UNIT

13.3

NAME __

COURSE __

DATE __

DRAWING

13.7

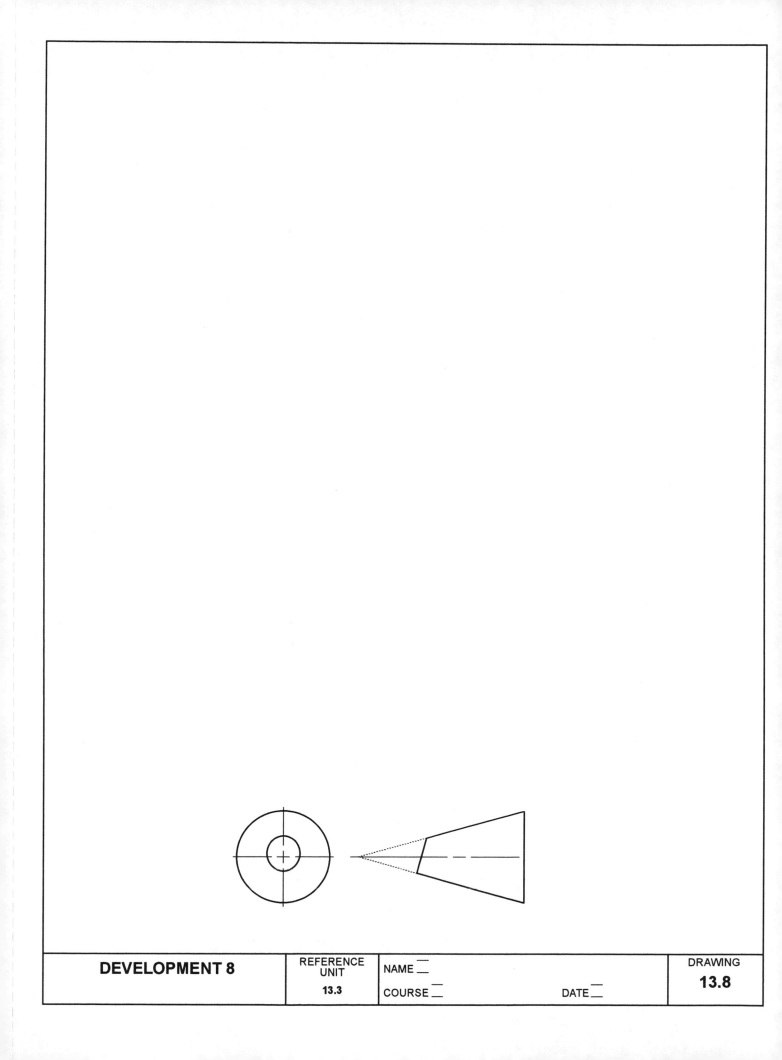

DEVELOPMENT 8

REFERENCE
UNIT

13.3

NAME __

COURSE __

DATE __

DRAWING

13.8

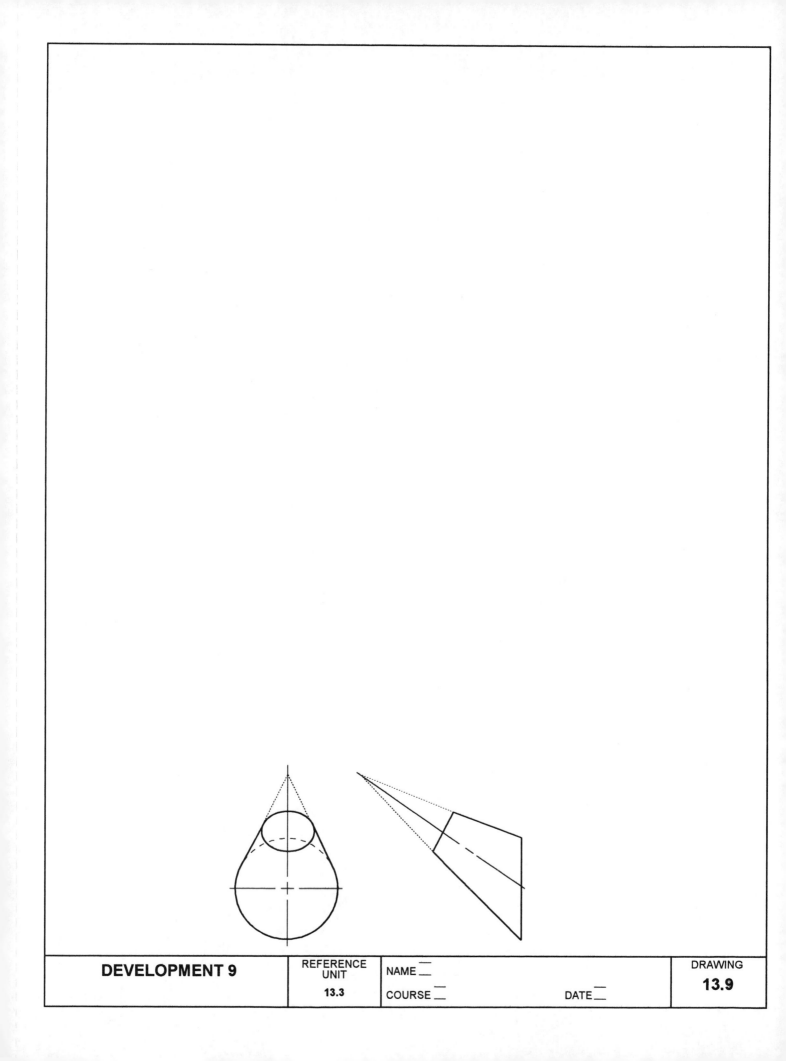

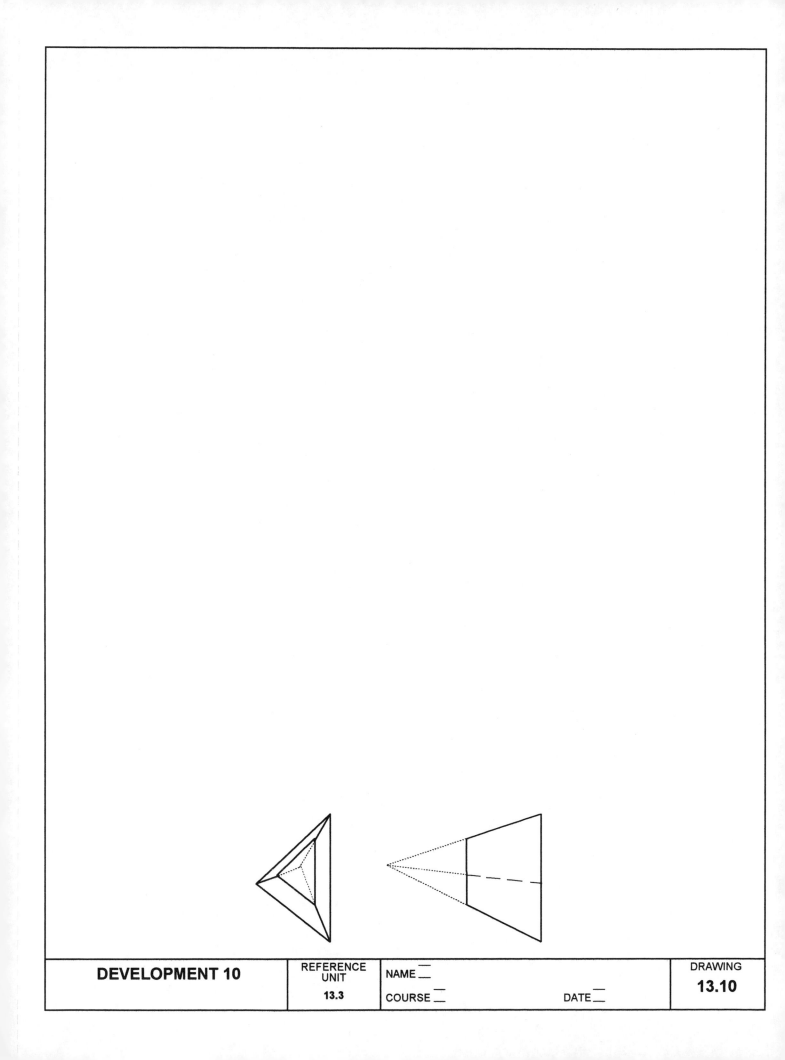

DEVELOPMENT 11

REFERENCE
UNIT

13.3

NAME —

COURSE —

DATE —

DRAWING

13.11

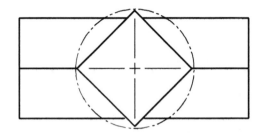

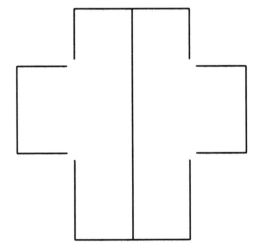

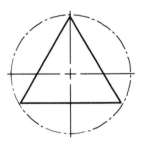

INTERSECTIONS 1	REFERENCE UNIT	NAME __		DRAWING
	13.2	COURSE __	DATE __	13.12

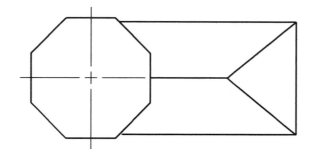

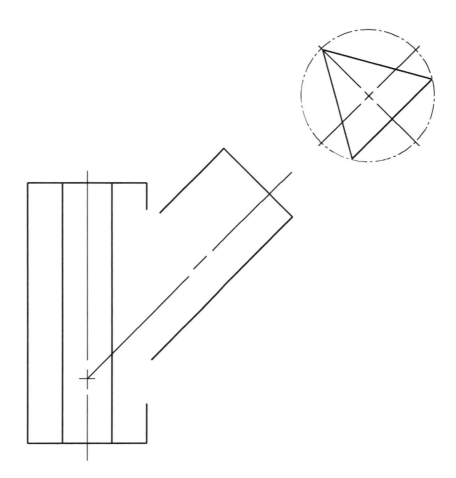

INTERSECTIONS 2

REFERENCE
UNIT

13.2

NAME ___

COURSE ___   DATE ___

DRAWING

13.13

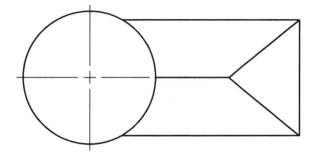

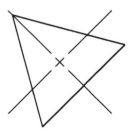

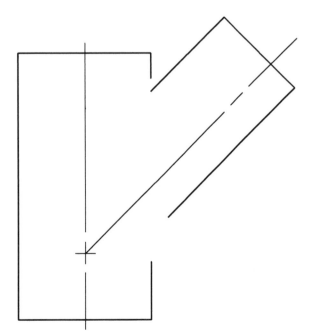

INTERSECTIONS 3

REFERENCE UNIT

13.2

NAME __

COURSE __

DATE __

DRAWING

13.14

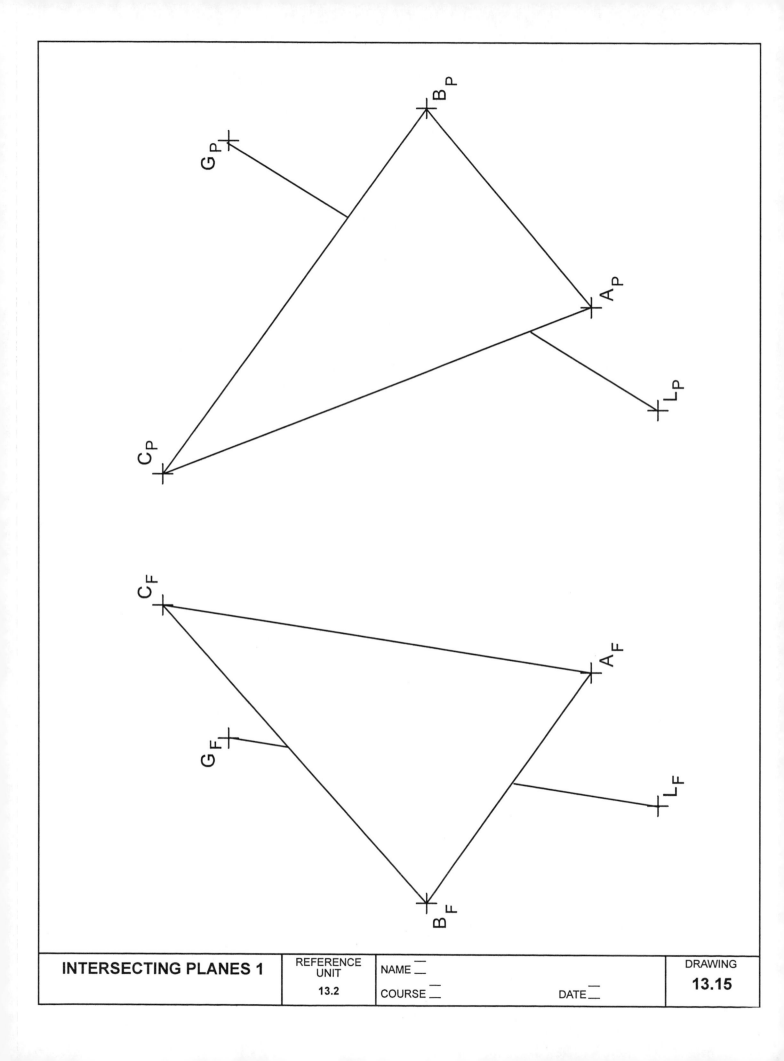

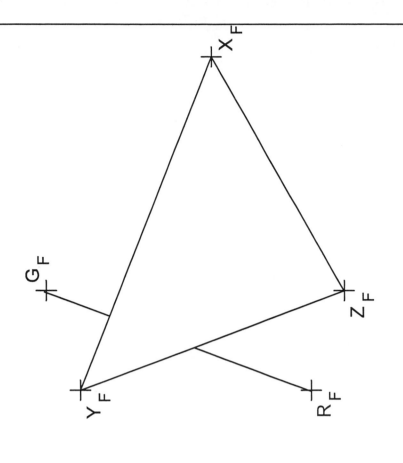

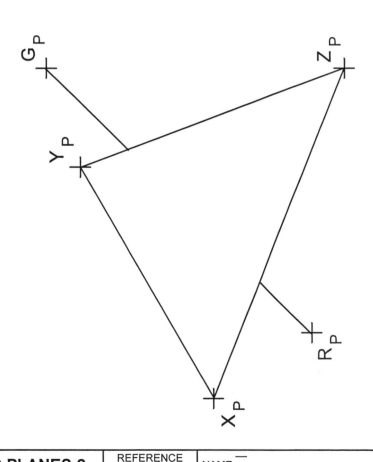

INTERSECTING PLANES 2	REFERENCE UNIT	NAME __		DRAWING
	13.2	COURSE __ DATE __		13.16

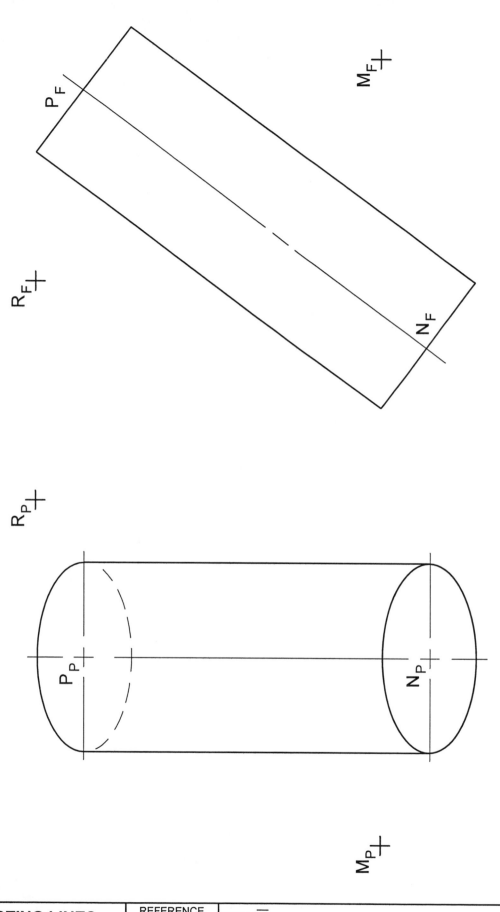

| INTERSECTING LINES AND PLANES | REFERENCE UNIT 13.2 | NAME __ COURSE __ DATE __ | DRAWING 13.17 |

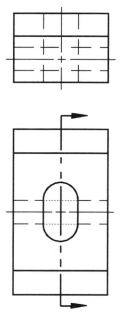

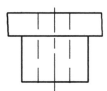

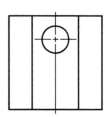

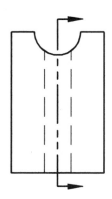

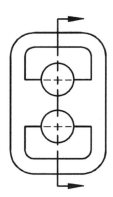

| SECTION VIEWS 1 | REFERENCE UNIT 14.4 | NAME ___ COURSE ___ DATE ___ | DRAWING 14.1 |

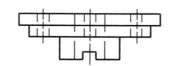

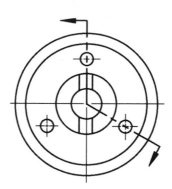

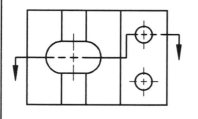

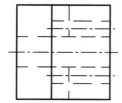

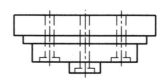

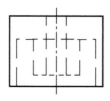

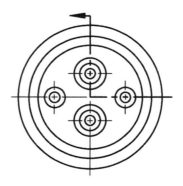

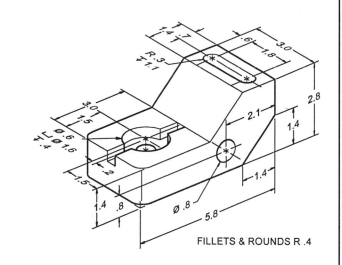

FILLETS & ROUNDS R .4

COUNTER BLOCK

REFERENCE UNIT

14.4

NAME ___

COURSE ___   DATE ___

DRAWING

14.3

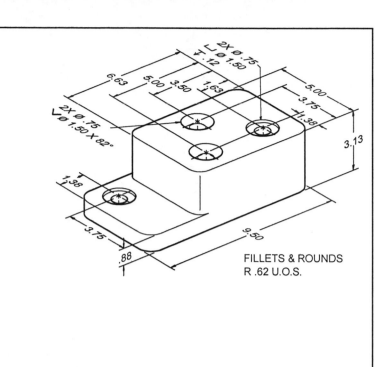

2X ⌀ .75
⌀ 1.50

2X ⌀ .75
⌀ 1.50 X 82°

6.63
5.00
3.50
1.63
5.00
3.75
1.38
1.38
3.75
9.50
.88
3.13

FILLETS & ROUNDS
R .62 U.O.S.

BRACKET	REFERENCE UNIT	NAME __		DRAWING
	14.4	COURSE __	DATE __	14.4

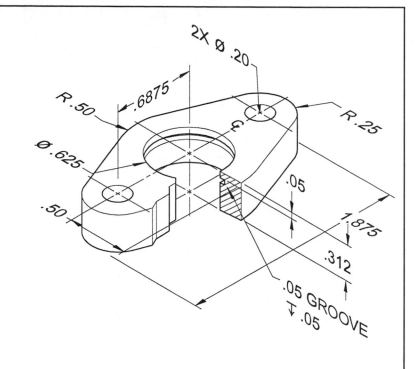

2X Ø .20

.6875

R .50

Ø .625

R .25

.05

.50

1.875

.312

.05 GROOVE
↧ .05

| RING COLLAR | REFERENCE UNIT 14.4 | NAME — COURSE — | DATE — | DRAWING 14.5 |

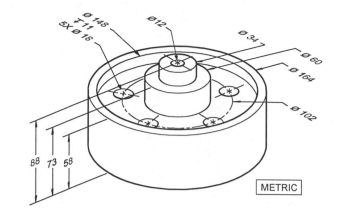

Ø148
▼11
5X Ø18

Ø12

Ø34

Ø60

Ø164

Ø102

88 73 58

METRIC

AXLE CENTER	REFERENCE UNIT	NAME __		DRAWING
	14.4	COURSE __	DATE __	14.6

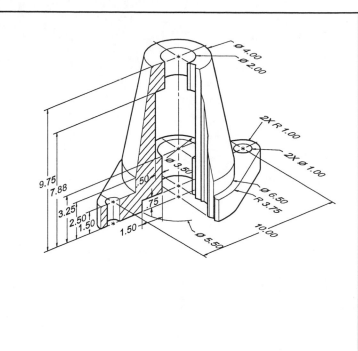

| TAPER COLLAR | REFERENCE UNIT 14.4 | NAME __ COURSE __ DATE __ | DRAWING 14.7 |

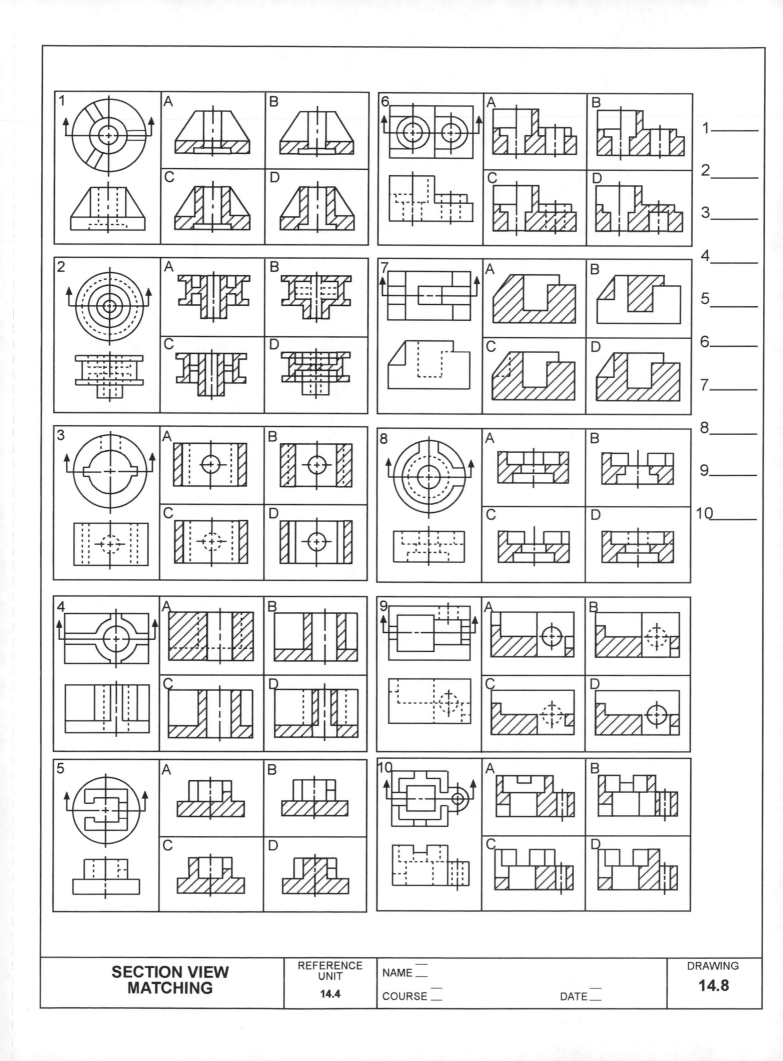

SECTION VIEW MATCHING

REFERENCE UNIT

14.4

NAME ___

COURSE ___

DATE ___

DRAWING

14.8

A

B

C

D

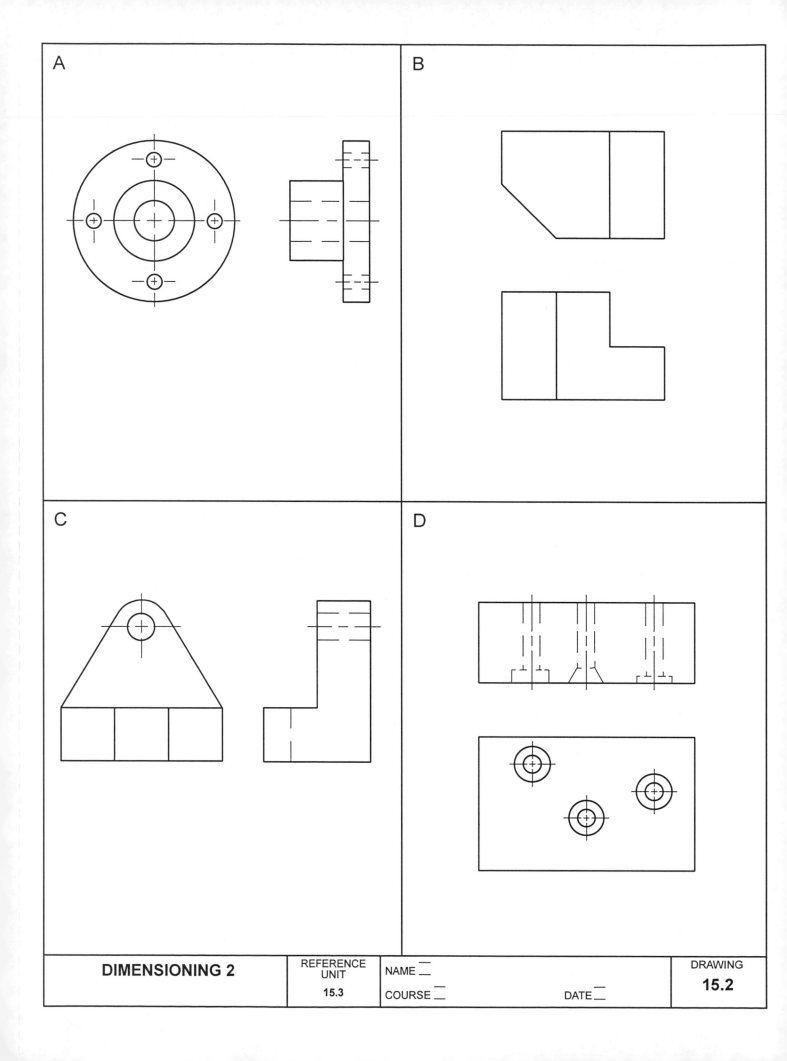

DIMENSIONING 2

REFERENCE UNIT

15.3

NAME —

COURSE —

DATE —

DRAWING

15.2

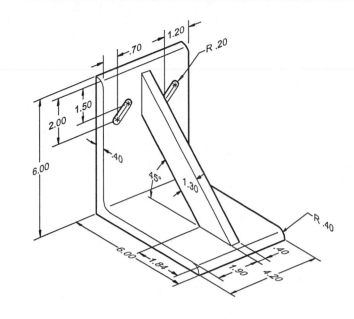

ANGLE BRACKET	REFERENCE UNIT	NAME __		DRAWING
	15.3	COURSE __	DATE __	15.3

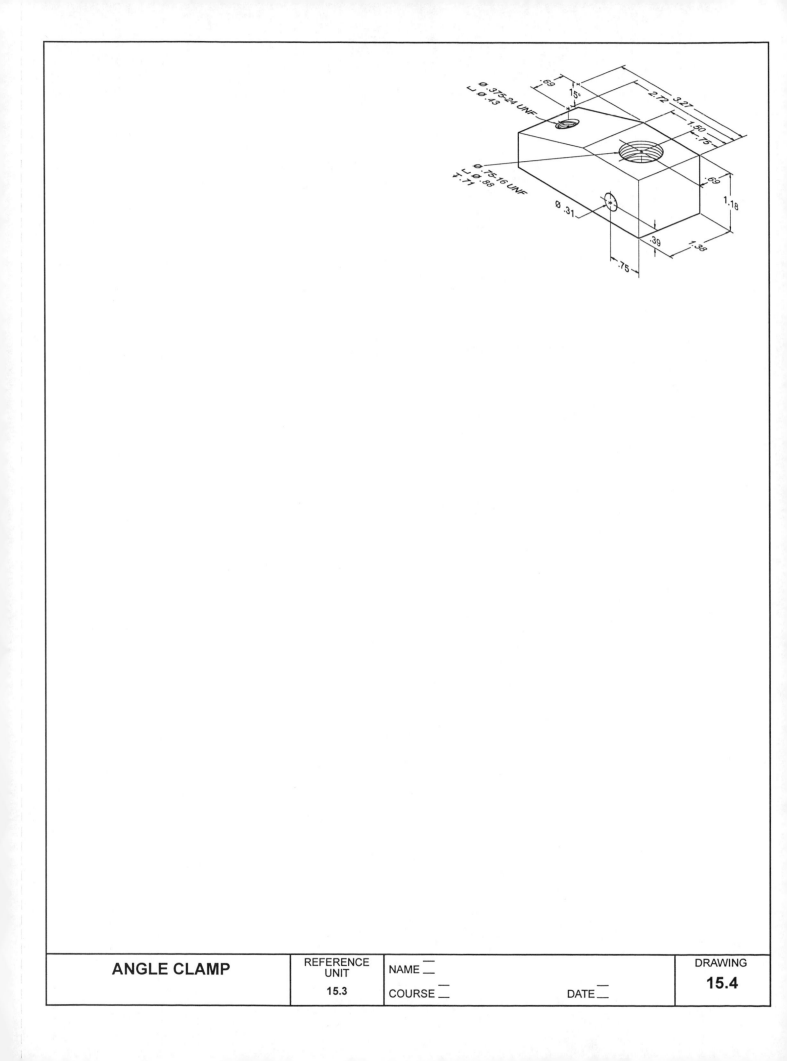

Ø .375-24 UNF
⌴ Ø .43

Ø .75-16 UNF
⌴ Ø .88
▼ .71

Ø .31

.69

15°

2.72

3.27

1.50

.75

.69

1.18

.39

1.38

.75

| ANGLE CLAMP | REFERENCE UNIT 15.3 | NAME —— COURSE —— DATE —— | DRAWING 15.4 |

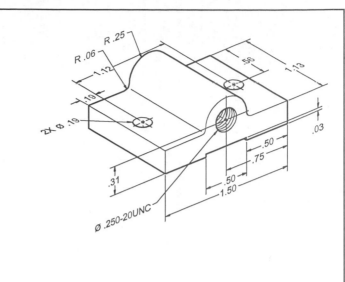

R .25
R .06
1.12
.19
.56
.13
2X Ø .19
.03
.31
Ø .250-20UNC
.50
.75
.50
1.50

OFFSET STRAP	REFERENCE UNIT	NAME —	DRAWING
	15.3	COURSE — DATE —	15.5

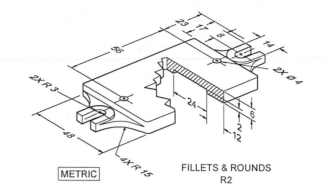

2X R 3

56

23
17
8

14

2X Ø 4

METRIC

4X R 15

48

24

6
2
12

FILLETS & ROUNDS
R2

MILL FIXTURE BASE	REFERENCE UNIT	NAME __		DRAWING
	15.3	COURSE __	DATE __	15.6

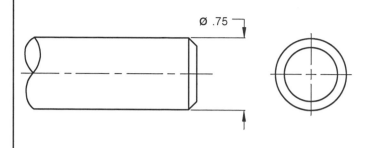
Ø .75
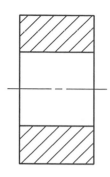
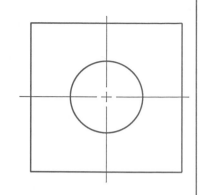

		Class of fit		RC 4		LC 6		FN 5
Hole		Nominal size						
	±	Limit	±		±		±	
	=	Upper limit	=		=		=	
		Nominal size						
	±	Limit	±		±		±	
	=	Lower limit	=		=		=	
Shaft		Nominal size						
	±	Limit	±		±		±	
	=	Upper limit	=		=		=	
		Nominal size						
	±	Limit	±		±		±	
	=	Lower limit	=		=		=	
Limits of fit		Smallest hole						
	−	Largest shaft	−		−		−	
	=	Tightest fit	=		=		=	
		Largest hole						
	−	Smallest shaft	−		−		−	
	=	Loosest fit	=		=		=	

TOLERANCING 1	REFERENCE UNIT 15.6.12	NAME __ COURSE __	DATE __	DRAWING **15.7**

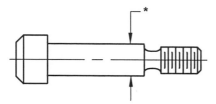

*See appendix for American Standard
Socket-head Shoulder Screw

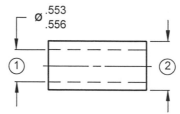

Bushing

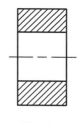

Housing

Limits of size			1		2 - FN 4 fit	
Hole		Nominal size		.500		.750
	±	Limit	±		±	
	=	Upper limit	=		=	
		Nominal size		.500		.750
	±	Limit	±		±	
	=	Lower limit	=		=	
Shaft		Nominal size		.500		.750
	±	Limit	±		±	
	=	Upper limit	=		=	
		Nominal size		.500		.750
	±	Limit	±		±	
	=	Lower limit	=		=	
Limits of fit		Smallest hole				
	−	Largest shaft	−		−	
	=	Tightest fit	=		=	
		Largest hole				
	−	Smallest shaft	−		−	
	=	Loosest fit	=		=	

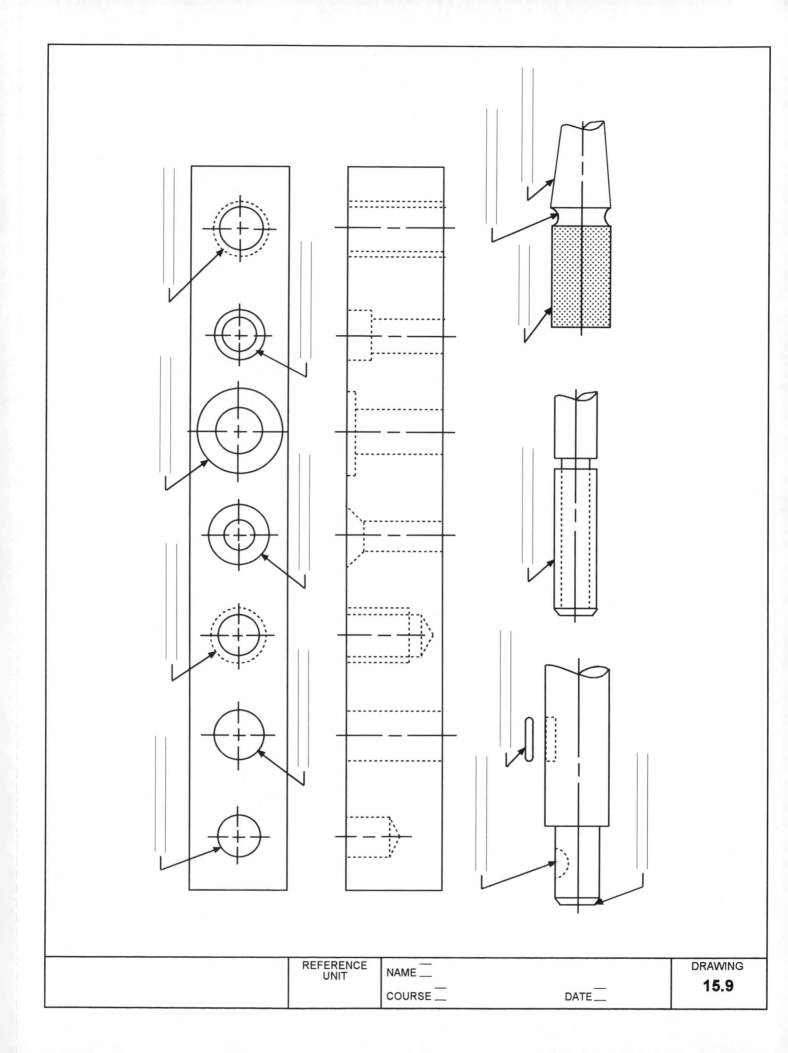

REFERENCE
UNIT

NAME___

COURSE___

DATE___

DRAWING

15.9

Geometric Characteristic Symbols

	Type of Tolerance	Characteristic	Symbol	See:
For Individual Features	Form	Straightness		6.4.1
		Flatness		6.4.2
		Circularity (Roundness)		6.4.3
		Cylindricity		6.4.4
For Individual or Related Features	Profile	Profile of a line		6.5.2 (b)
		Profile of a surface		6.5.2 (a)
For Related Features	Orientation	Angularity		6.6.2
		Perpendicularity		6.6.4
		Parallelism		6.6.3
	Location	Position		5.2
		Concentricity		5.11.3
		Symmetry		5.13
	Runout	Circular Runout		6.7.1.2.1
		Total Runout		6.7.1.2.2

* Arrowheads may be filled or not filled

1. Surface B must be parallel to datum surface A within 0.10 mm.
2. The axis of each hole must lie within a tolerance zone of 0.03 mm diameter.
3. The datum surface A must be flat within 0.03 mm.

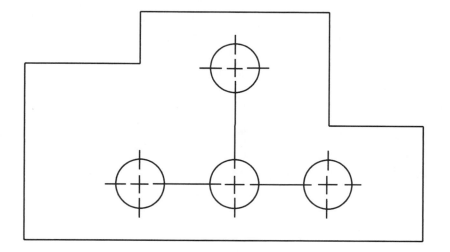

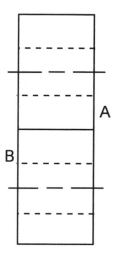

1. The small hole must be concentric with cylinder C within 0.010 mm.
2. Cylinders A and B are to have a cylindricity tolerance of 0.03 mm.
   Cylincer C is the datum.

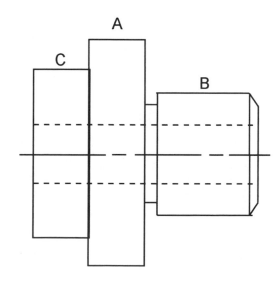

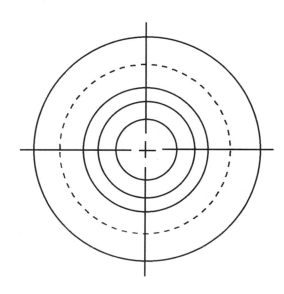

GEOMETRIC TOLERANCING 2	REFERENCE UNIT 16.2	NAME __ COURSE __	DATE __	DRAWING 16.2

1. Surfaces C and D must be perpendicular to datum surface E within 0.03 mm.
2. The vertical hole B must be perpendicular to datum hole A within a tolerance of 0.05 mm.

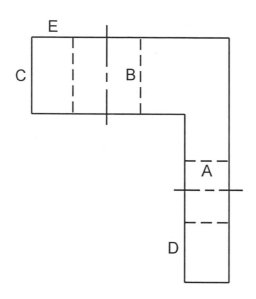

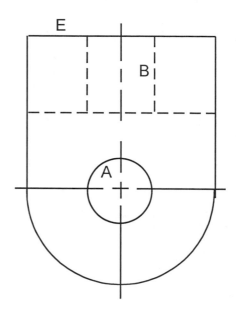

1. Surfaces B and C must be perpendicular to datum surface A within 0.03 mm.
2. The axis of the hole must be parallel to datum surface A within 0.05 mm.

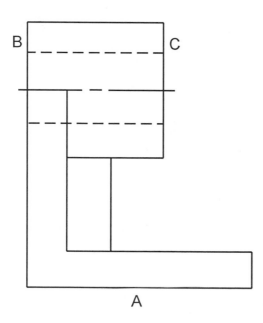

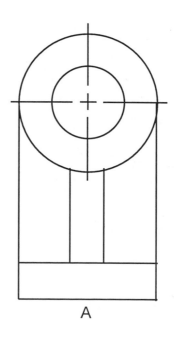

| GEOMETRIC TOLERANCING 3 | REFERENCE UNIT 16.2 | NAME __ COURSE __ | DATE __ | DRAWING 16.3 |

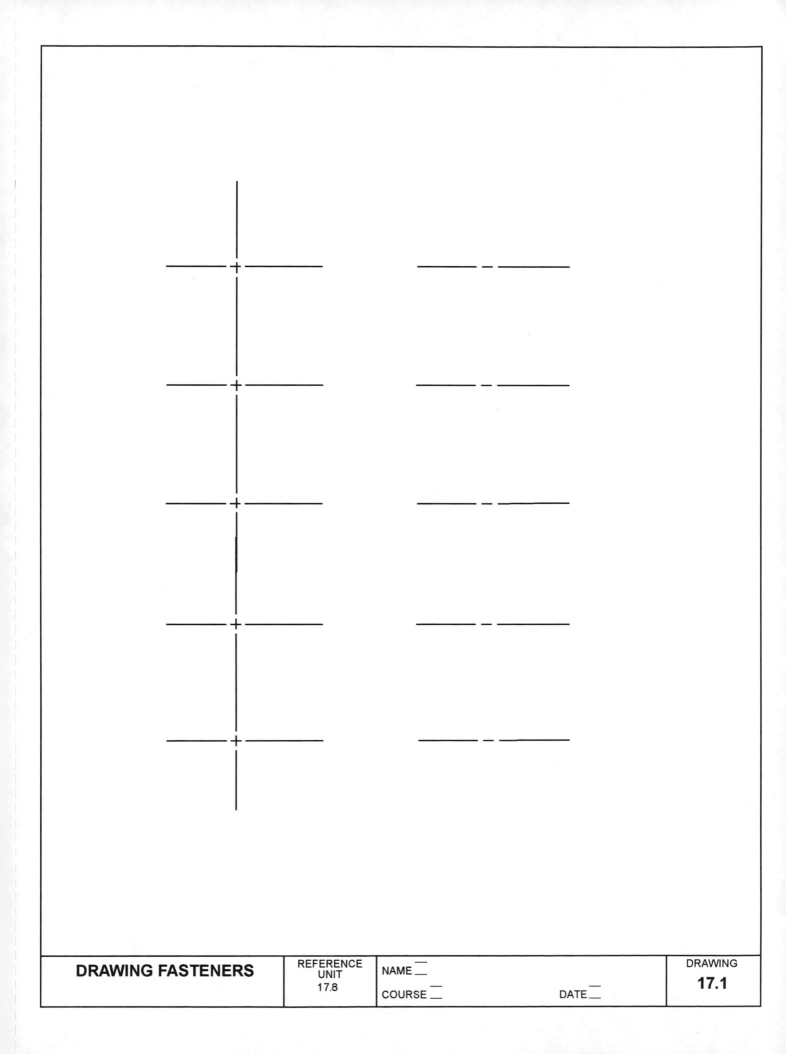

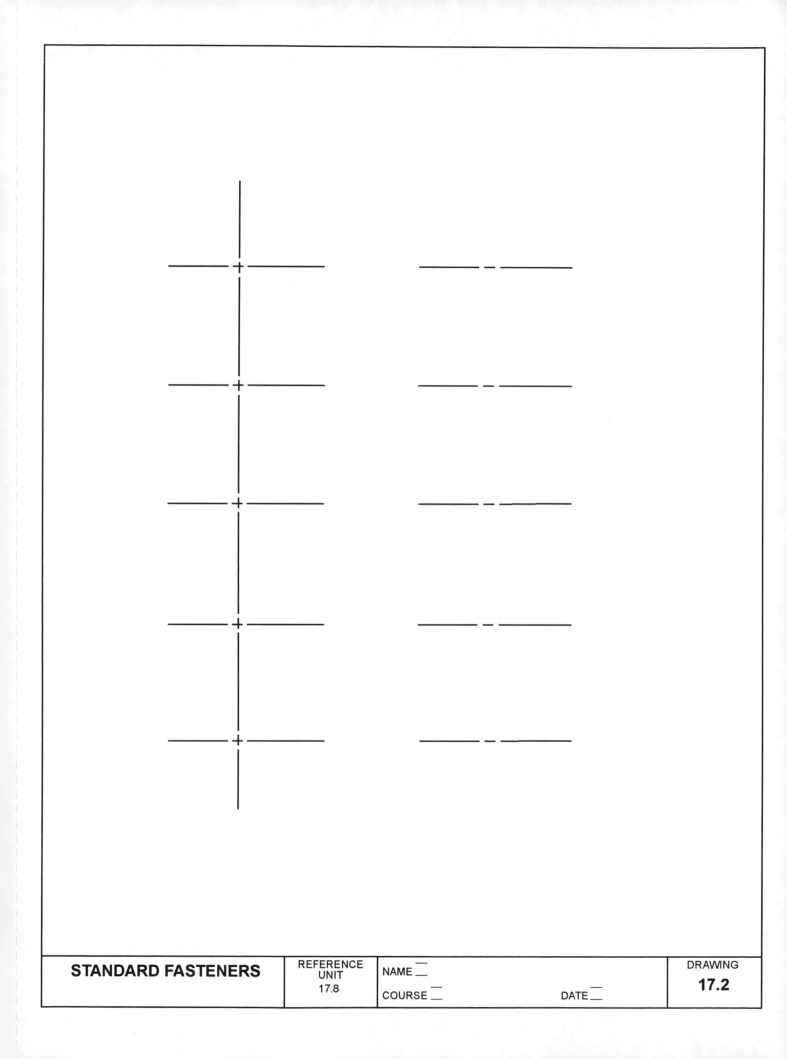

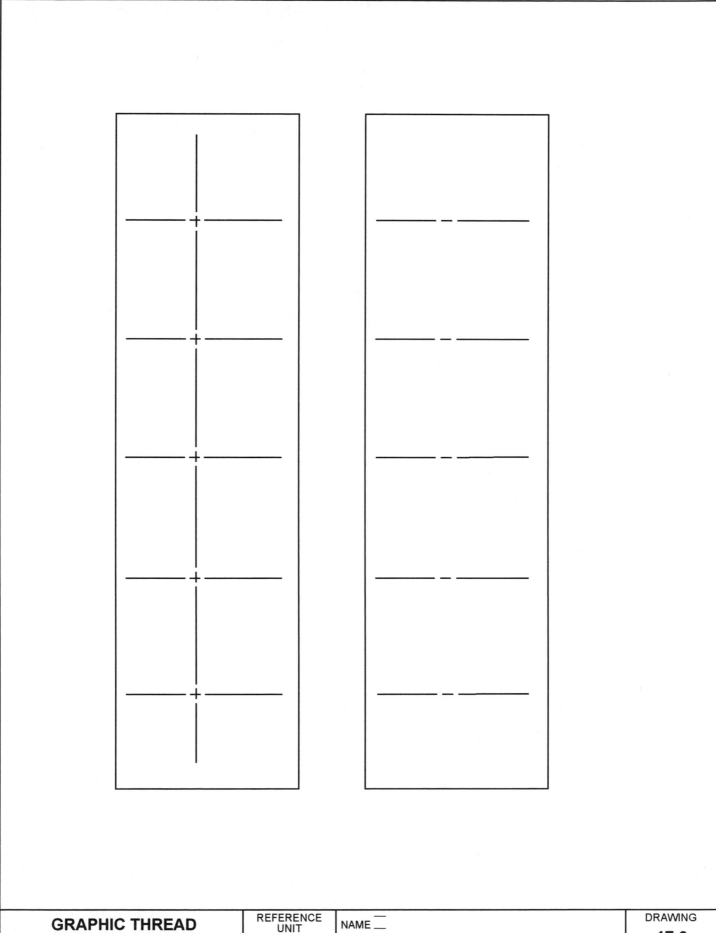

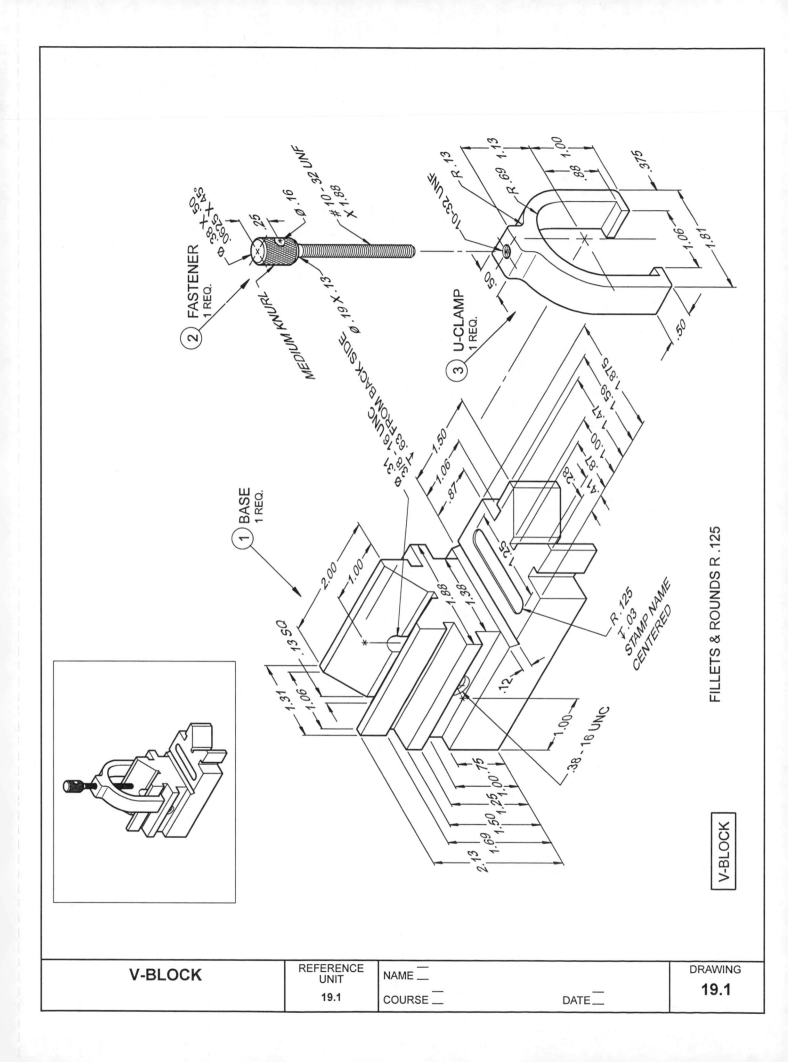

FILLETS & ROUNDS R .125

V-BLOCK

| V-BLOCK | REFERENCE UNIT 19.1 | NAME __ COURSE __ DATE __ | DRAWING 19.1 |

| | V-BLOCK | REFERENCE UNIT 19.1 | NAME __ COURSE __ DATE __ | DRAWING 19.1 |

| V-BLOCK | REFERENCE UNIT 19.1 | NAME __ COURSE __ DATE __ | DRAWING 19.1 |

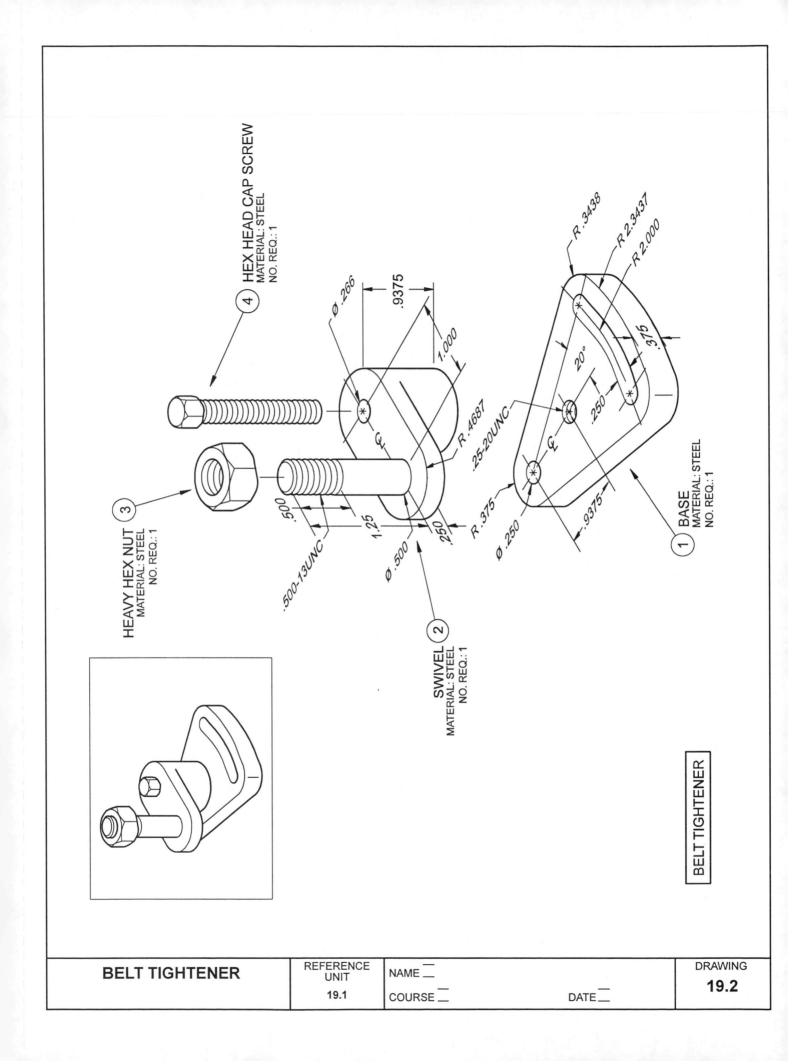

HEX HEAD CAP SCREW
MATERIAL: STEEL
NO. REQ.:1

(4)

HEAVY HEX NUT
MATERIAL: STEEL
NO. REQ.:1

(3)

Ø .266

.9375

1.000

R .4687

.25-20UNC

.500

.500-13UNC

1.25

Ø .500

.250

SWIVEL
MATERIAL: STEEL
NO. REQ.:1

(2)

R .3438

R 2.3437

R 2.000

20°

.375

.250

R .375

Ø .250

.9375

BASE
MATERIAL: STEEL
NO. REQ.:1

(1)

BELT TIGHTENER

| BELT TIGHTENER | REFERENCE UNIT | NAME __ | DRAWING |
| | **19.1** | COURSE __ DATE __ | **19.2** |

Heat of Combustion of Fuel Sources

Substance	BTU/lb
Acetylene	21390
Alcohol, methyl	9560
Benzine	18140
Charcoal, wood	13440
Coal, bituminous	12780
Gasoline	21350
Kerosene	19930
Hardwood	6980

BAR GRAPH	REFERENCE UNIT 20.3.1	NAME COURSE DATE	DRAWING 20.1

Number of Teeth	Diametral Pitch	Involute
12	4	14.5 degrees

	SPUR GEAR	REFERENCE UNIT	NAME __		DRAWING
		22.2.9	COURSE __	DATE __	22.1

Cam type	Face
Base circle	4"
Roller type	Knife
Roller diameter	0.625"
Shaft diameter	0.75"
Hub diameter	1"
Rotation direction	Clockwise
Follower position	Vertical over center of the base circle
Plate thickness	0.5"
Rise	1.25"
Motion description	Rises with uniform acceleration for 180 degrees, then drops with uniform acceleration for 180 degrees.

CAM PROFILE

REFERENCE UNIT

22.3.5

NAME __

COURSE __

DATE __

DRAWING

22.2

NAME

COURSE

DATE

NAME

COURSE   DATE

NAME

COURSE   DATE

NAME

COURSE

DATE

DRAWING

DRAWING

NAME

COURSE

DATE

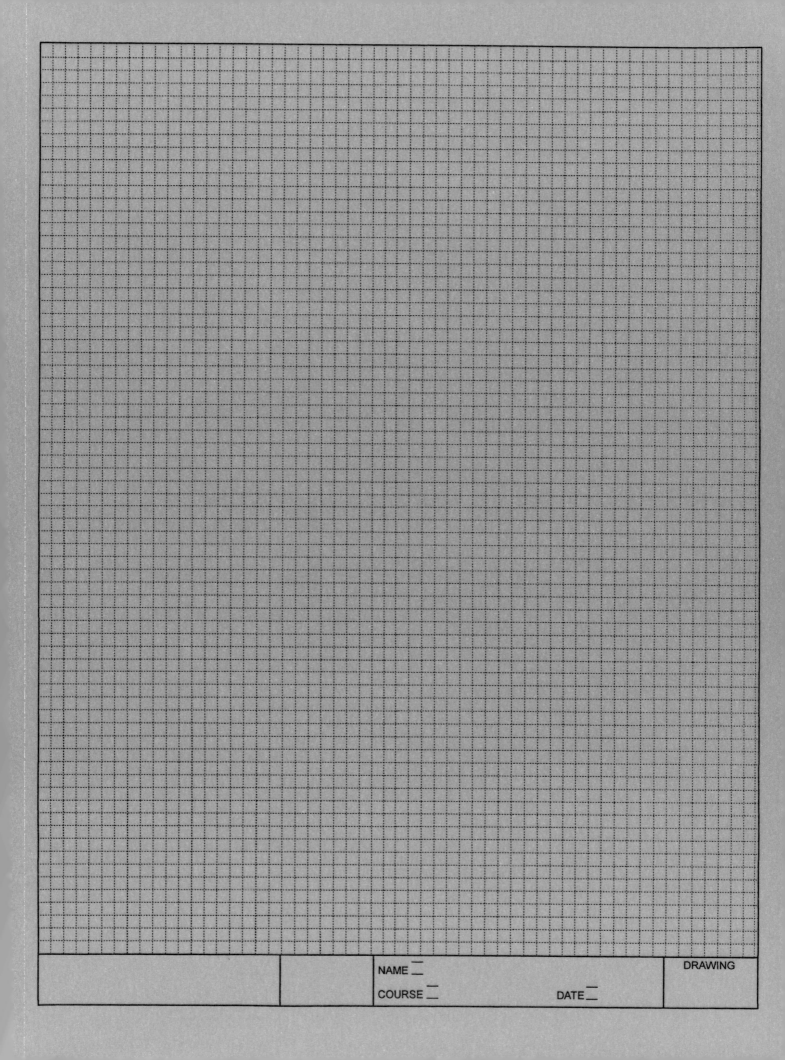

NAME ‗

COURSE ‗

DATE ‗

DRAWING

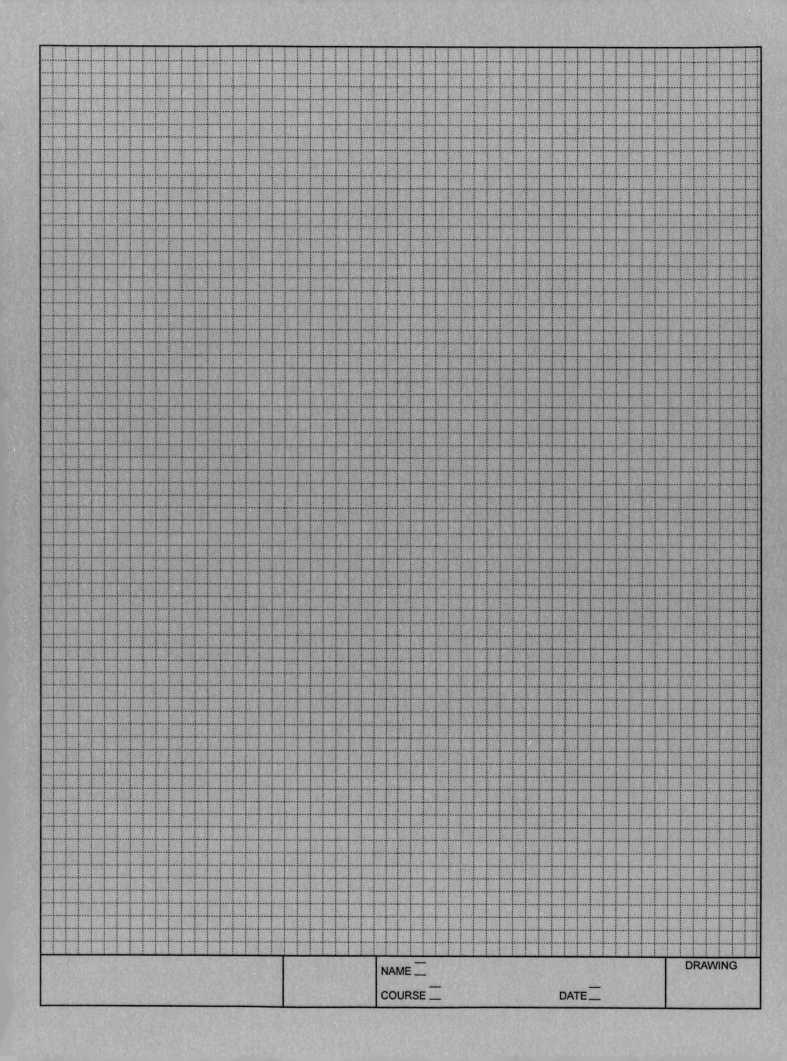

NAME ‾‾
COURSE ‾‾   DATE ‾‾

DRAWING

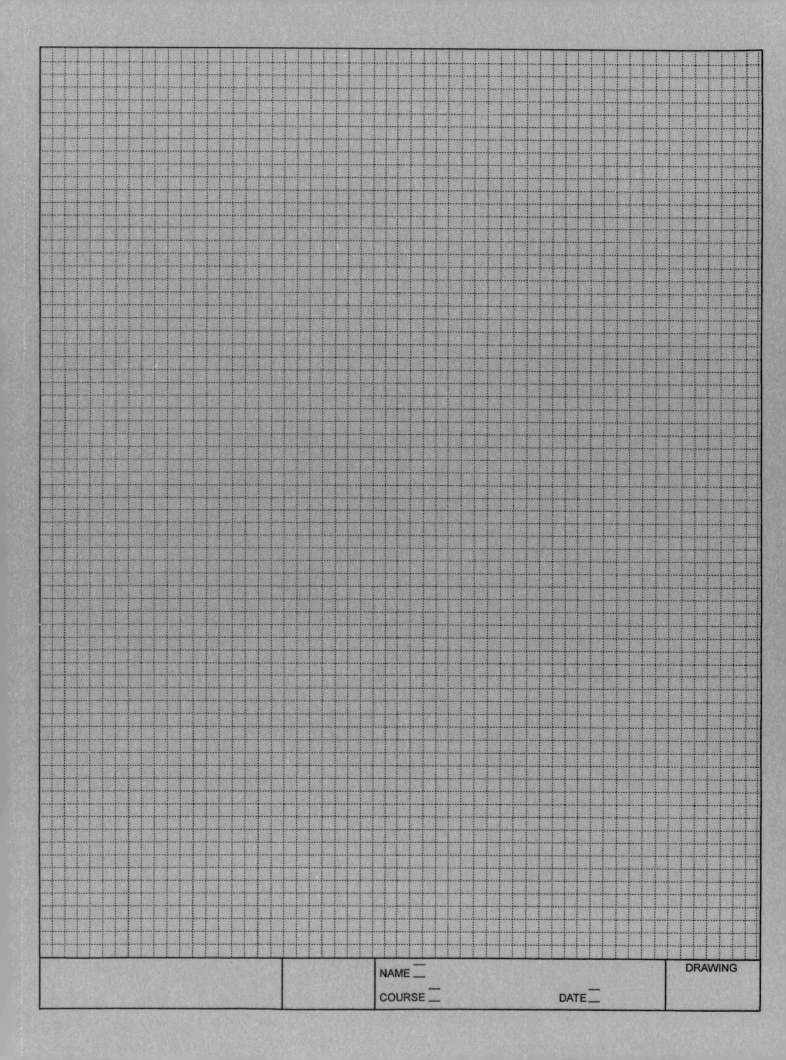

NAME __

COURSE __   DATE __

DRAWING

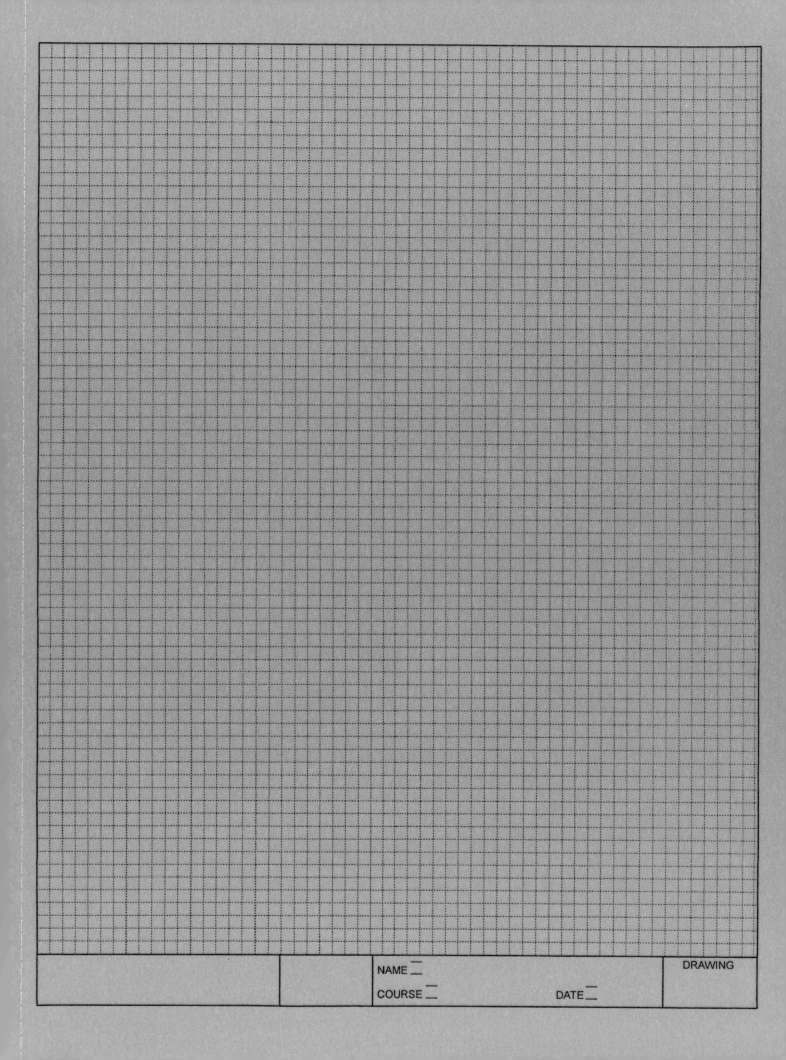

NAME __

COURSE __   DATE __

DRAWING

NAME __

COURSE __

DATE __

DRAWING

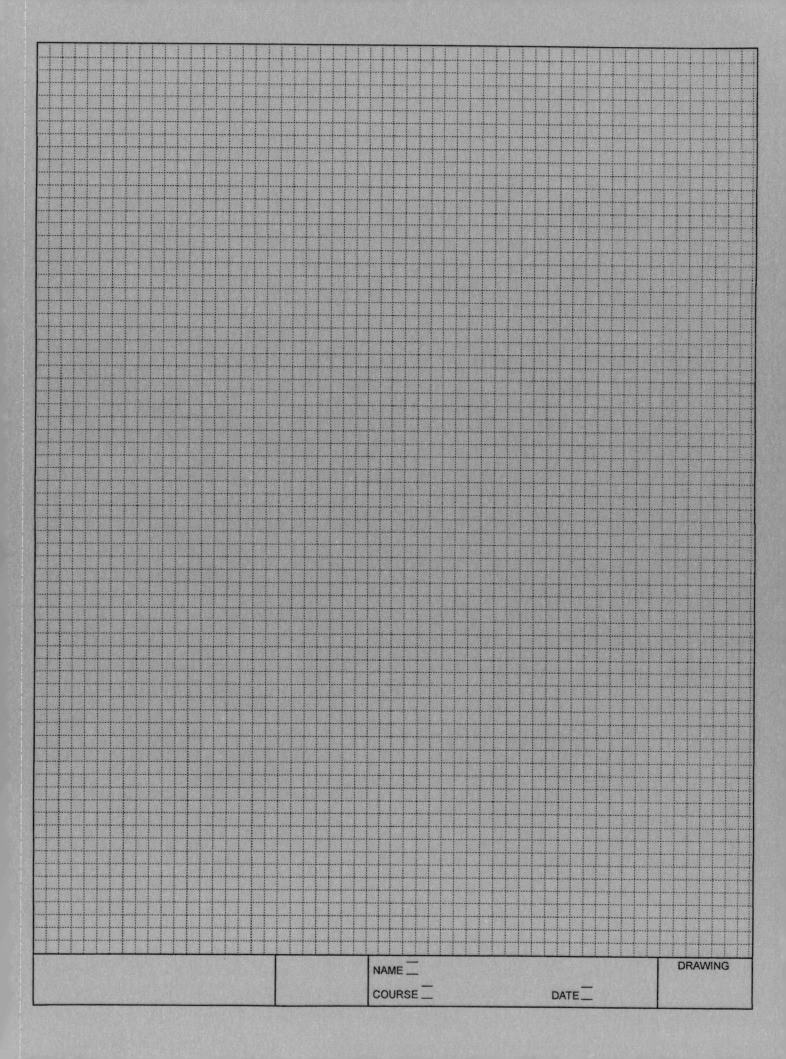

NAME __

COURSE __

DATE __

DRAWING

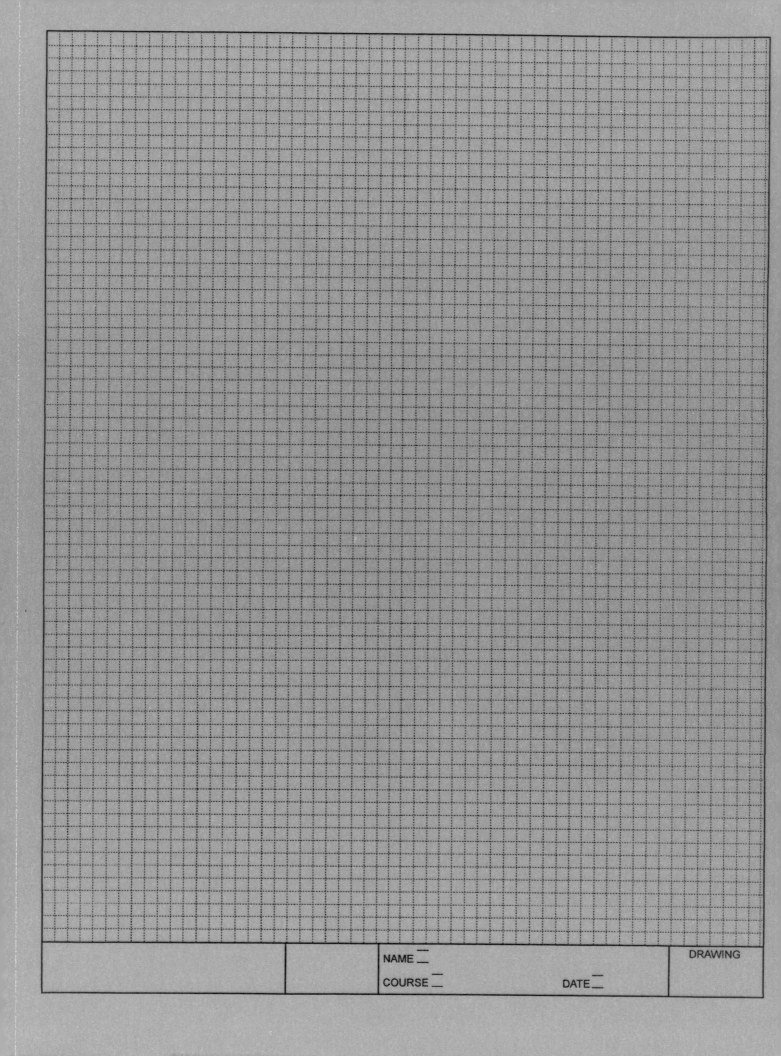

NAME __

COURSE __   DATE __

DRAWING

DRAWING

NAME

DATE

COURSE

NAME __

COURSE __   DATE __

DRAWING

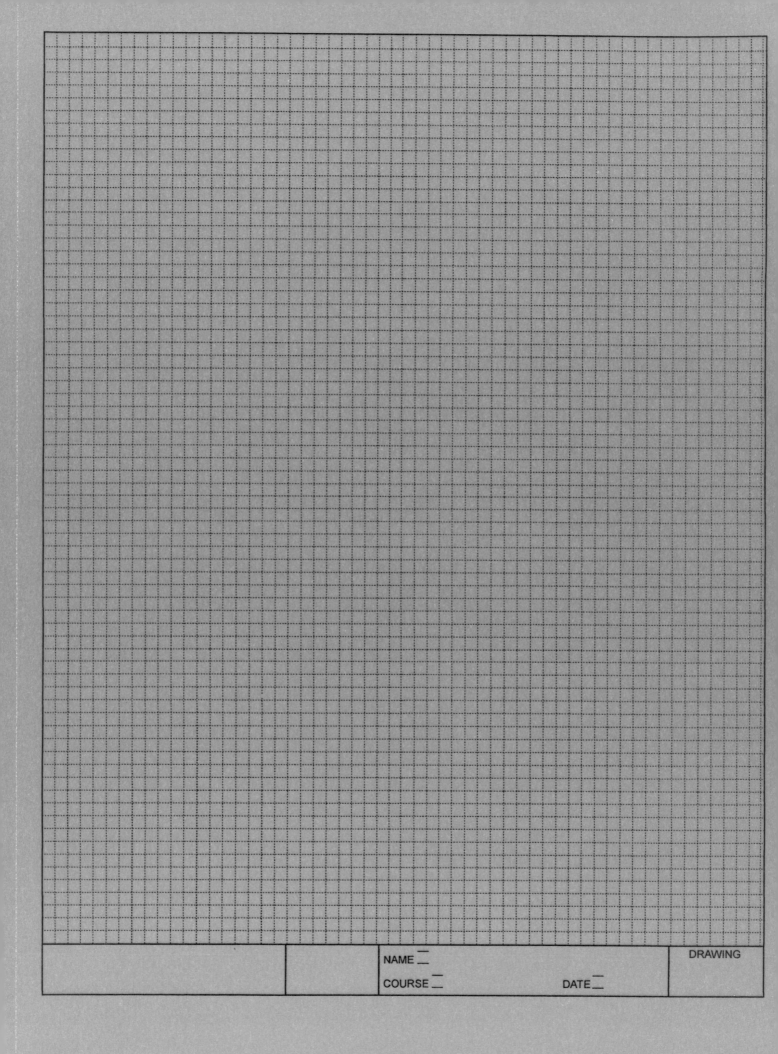

NAME __

COURSE __   DATE __

DRAWING

NAME __

COURSE __   DATE __

DRAWING

NAME __

COURSE __　　　DATE __

DRAWING

NAME __

COURSE __

DATE __

DRAWING

DRAWING

NAME ___

COURSE ___

DATE ___

NAME __

COURSE __   DATE __

DRAWING

NAME __

COURSE __   DATE __

DRAWING

DRAWING

NAME ___

COURSE ___ DATE ___

DRAWING

NAME ____

DATE ____

COURSE ____

NAME __

COURSE __

DATE __

DRAWING

NAME __

COURSE __

DATE __

DRAWING

DRAWING

NAME ‾

COURSE ‾   DATE ‾

NAME __

COURSE __   DATE __

DRAWING

DRAWING

NAME

COURSE

DATE

NAME __

COURSE __   DATE __

DRAWING

DRAWING

NAME ▭

COURSE ▭   DATE ▭

NAME __

COURSE __   DATE __

DRAWING

DRAWING

NAME___

COURSE___

DATE___

NAME __

COURSE __

DATE __

DRAWING